Deserted Devices and Wasted Fences

Deserted Devices and Wasted Fences

everyday technologies in extreme circumstances

Dani Ploeger

Published in this first edition in 2021 by:
Triarchy Press
Axminster, England

info@triarchypress.net
www.triarchypress.net

A catalogue record for this book is available from the British Library.

Print ISBN: 978-1-913743-43-7
ePub ISBN: 978-1-913743-44-4

Printed in the UK by TJ Books, Padstow, Cornwall

Contents

Introduction 1

1. Tactical Transgressions: Bashar al-Assad's phone 4

2. E-Waste in Cling Film: The symbolic order of technological progress 8

3. Hi-Tech Everything: A report from the heart of techno-consumerism 14

4. Eerie Prostheses and Kinky Strap-Ons: Mori's uncanny valley and ableist ideology 20

5. The Dirt Inside: Computers and the performance of dust 28

6. Orodha: The ultimate fetish commodity and its reversal 34

7. Frugal Phone / Material Medium 39

8. Positioning the Middle of Nowhere: GPS technology and the desert 42

9. Sounds of Violence: The affective tonality of high-tech warfare 48

10. Smart Bombs, Bulldozers and the Technology of Hidden Destruction 53

11. Smart Technologies and Soviet Guns: The dialectics of postdigital warfare 60

12. Techno-Mythology on the Border: The pandemic risk society........ 69

13. Camera Surveillance and Barbed Wire........ 75

14. The Smart Fence is the Message: EU border barriers as violent media 78

15. The Deluxe Anti-Terrorist Barrier 83

16. Struggle and Expand: The Delta Works as colonial technology........ 86

Postscript: Artificial techno-myths 89

Acknowledgements 95

References........ 105

About the author........ 111

Introduction

In the mid-2000s, just after the mobile phone boom had happened, I spent two years living in Ramallah, Palestine. While the conflict between the newly elected Hamas government and the State of Israel played out, it struck me how under these tense circumstances, the everyday roles of mobile technologies often differed from their common use and representation in the Northern European context I was used to. Mobile phones were a crucial way to warn of military incursions, while they were also treated with suspicion because of their potential appropriation as bomb triggers. Meanwhile, the phone model I owned was colloquially named after Syria's president Bashar al-Assad, and thus became politically charged.

Since that time, I have been interested in the ways in which everyday technologies are used and appropriated under circumstances that their designers and manufacturers did not envisage and in ways that do not correspond with their usual representation in advertising and other media. In these situations, the taken-for-granted understandings of technology that prevail in the 'ordinary everyday' of consumer culture in the so-called Global North are often disturbed or contradicted and consumer devices no longer appear as the unambiguous signifiers of innovation and progress they are often taken to be. I like to look at these ruptures as opportunities to imagine alternative techno-cultures, in which technologies are developed, used and appropriated in diverse ways that are rooted in people's personal, local and regional lifeworlds, rather than act as extensions of globalized standardization processes.

As such, my approach bears a certain kinship with Henri Lefebvre's call (1987) to study occasions of 'festival' ('la fête') in society to look for remnants of medieval elements that are 'other' to the everyday of consumer capitalism. These festive elements can be a source to help us imagine and initiate a revolutionized everyday life that breaks free from a culture of passivity, both with regard to its alienating labour processes and the endless consumption of standardized goods. However, instead of looking for moments of disruption at instances of festival that take place at the heart of consumerism, I suggest that we look at the opposite end of the spectrum, namely at the fringes of consumer culture, in places that are on the "frontiers of capitalism" as Raj Patel and Jason W. Moore (2017) have called the geographical and cultural areas in which the mechanisms of capitalist production and consumption are not yet – or no longer – fully established. Examining the ways in which technologies that were designed, produced and marketed for use in everyday consumer culture start to (mal)function, gain new meanings and are appropriated in these liminal spaces can give us hints at what technology beyond global consumerism could look like.

Over the years, I have travelled through Europe, Africa and Asia to examine such situations of technological rupture and create artwork in response to this. I have visited e-waste dump sites in Lagos, Nigeria, second-hand markets and mobile phone shops across Kenya, Pashtun migrant areas on the outskirts of Karachi, the frontline of the Donbass War in East Ukraine, the Polish-German border during the Covid-19 crisis, an e-waste recycling plant in Hong Kong, the anti-immigration 'smart fence' on the Hungarian border, sex shops in London's Soho, a consumer technology trade fair in Berlin and many other sites of technological turmoil, disruption and surprising convergences.

These journeys have formed the basis for my artistic practice around consumer devices in relation to waste, conflict and intimacy over the past decade, but they have also led to a series of short texts that reflect on everyday technologies, their uses and imaginaries, most of which have thus far remained unpublished. This collection brings together a selection of these texts, written between 2013 and 2020. The persistence of diverse experiences of materiality – despite efforts for technological immediacy – is a recurring theme throughout them, as are the examined technologies' implication in the dissemination of ideologies of progress, growth and conquest.

My motivation to share these texts is not to offer a handbook for cultural change based on conclusive analyses – most of the essays are open-ended reflections – but rather to offer some provocations, which I hope might stimulate alternative perspectives on the technologies we often overlook or take for granted in everyday life. I hope that my guided tour along the dusty insides of computers, sharp edges of razor wire technology, large-yet-not-uncanny strap-on dildos and other technological curiosities and banalities might trigger some doubts about the techno-cultural status quo and evoke a desire for more inclusive technologized lifeworlds.

Nairobi, November 2020

1. Tactical Transgressions:

Bashar al-Assad's phone

In the summer of 2006 I lost another phone in a taxi in Ramallah. On the corner of Manara Square in the centre of town I bought a second-hand Nokia and a new prepaid SIM card. A colleague at the music school I worked at gave me a stern warning about my developing habit of losing phones. He claimed that somebody could use one of my lost phones to detonate a bomb, which would get me into trouble since the SIM card was registered in my name. To my knowledge such cases have never occurred and the story was more like an urban legend going around at the time, but the rumour didn't come from nowhere.

Even when bought on a street corner, pre-paid SIM cards could only be obtained upon presentation of a valid ID of which a copy was made by the sales person who subsequently had to submit this to the Palestinian Authority. The connection between the introduction of this regulation and the use of phones for bomb attacks that was often made informally was not far-fetched, since a mobile phone detonator was used in the 2002 bombing of The Hebrew University of Jerusalem, one of the first known cases of this practice (Boulden, 2004). The assumption was that the Israeli authorities, who have had far-reaching control over Palestinian telecommunications despite agreements on independence in the Oslo Accords (Worldbank,

2013), would be behind the ID registration requirements and thus have access to the data.

My new-second-hand Nokia 3310 was called Bashar, one of my students told me. When I asked him about the origins of this nickname, he said that Bashar al-Assad, the president of Syria, probably used to have this model when it came out. Intrigued by the practice of naming phones – my previous one turned out to be called 'timsah', or crocodile – I asked a few other people about the origins of Bashar's name. Nobody was certain but the reference to al-Assad occurred several times. This made sense, considering the narrative of liberalization and modernization that had been propagated about the new Syrian president who had succeeded his father Hafiz al-Assad at the beginning of the decade.

However, asking around about the phone naming question 15 years later, it quickly turns out that al-Assad might never have owned a Bashar and that the Nokia 3310's nickname actually has nothing to do with him. It originates in an advertising campaign run by telecom company TMC from 2000. In the ads, Kuwaiti star footballer Bashar Abdullah promoted the device. While a Kuwaiti pointed this out to me, none of the Palestinians I ask recall the advertisement. TMC is not active in the Palestinian territories, so the chances of exposure to the campaign with Bashar Abdullah would have been limited. It seems that on the fringes of the campaign's medial reach it seems that the origin myth of the Bashar nickname had started to live its own life.

While drawing on largely speculative and accidental information, this anecdotal account of some stories surrounding a mobile phone give an impression of the ways in which the intended uses and meanings of consumer technologies are diffracted and repurposed through everyday processes. A marketing campaign that appears aimed at positioning a new phone model as a politically neutral commodity by associating it with a famous football player accidentally leads to the device being perceived in the context of a media campaign surrounding a dictatorial ruler. While intended for telecommunication, mobile phones are hacked to repurpose them as bomb detonators. This, in turn, leads to speculative narratives of threat (your lost phone being used for terrorist attacks) that contribute to an environment of fear surrounding everyday consumer technology use.

Taking the everyday practices surrounding a commodity as a starting point for broader reflections on culture is an approach that underpins the work of thinkers such as Thorstein Veblen (1899), Henri Lefebvre (1987) and Michel de Certeau. Whereas social theory has traditionally focused on the realm of politics and production and often neglected the practices of acquiring goods and services, these theorists consider the everyday – that what remains when "all distinct, superior, specialized, structured" life activities are stripped away (Lefebvre, 1991 [1958]) – to be the primary domain of culture where commodities are remade as cultural artefacts and people shape their identities (Poster, 2004).

In this context, the development of everyday uses and meanings of consumer technologies can be read from the perspective of Michel de Certeau's (1984) theorization of the shaping of cultural practices on the basis of *strategies* and *tactics.* Entities who hold power over the production, distribution and regulation of technologies seek to control the ways they are used through the implementation of strategies. Devices and software are often designed and regulated in ways that stimulate their smooth operation as a consumer commodity. For example, electronic gadgets are designed to break or malfunction after a limited time to ensure regular purchase of new devices, while applications in the Apple App

Store are admitted on the basis of their adherence to strict user interface protocols and the principle that users should not be able to program the device themselves. These strategies are countered by some users through tactics. Various organizations and online communities promote work-arounds that can be exploited to repair devices that are designed to be replaced upon malfunctioning[1] and mobile devices are often 'rooted' to override manufacturer-determined software restrictions.

However, the anecdotes around the Bashar phone show that in day-to-day use the interplay between the strategies of entities in power and the countering tactics of those lacking power are not as clear cut and straightforward as de Certeau's model might suggest at first sight. There appears to be an intertwinement of strategies and tactics, where some tactics may simultaneously fit in with a strategic framework in another context. On one hand, manufacturers' use of model numbers that are reminiscent of sci-fi narratives is a way to position their devices in the high-tech realm. The nicknaming of mobile phones could be seen as a tactic to remove them from that ideological realm, with its connections to myths of progress and innovation (see e.g. Gray, 1999). By renaming Nokia 2650, 3310, 6600 as Crocodile, Bashar and Panda, the devices become associated with animate creatures rather than technological systems. On the other hand, the nicknaming of the Nokia 3310 as Bashar, while originating from a seemingly politically-neutral advertising campaign by the manufacturer or distributor, led to the device's association with the strategic domain of official propaganda around the Syrian president's image as a modernizer and liberal head of state.

---

[1]One such organization is the London-based social enterprise The Restart Project, which generates, collects and disseminates information and skills on how to repair domestic technologies.

2. E-Waste in Cling Film:

The symbolic order of technological progress

While we are searching for electronic waste originating from Europe around Lagos, Nigeria, Jelili and I mostly find fully functional desktop computers, monitors and laptops, cleanly wrapped in cling-film. In December 2013, I visit the so-called Computer Village in Ikeja, Lagos, one of Africa's largest ICT markets, together with Nigerian performance artist Jelili Atiku. In our effort to trace the afterlife of digital devices that have been discarded by their users in Europe, we first tried to find places where digital debris would be brought into the country. However, it quickly became clear that the vast majority of devices imported from Europe are actually still functional and in high demand on local second-hand markets. In Ikeja, we end up buying a neat 15" monitor, the sticker of its corporate previous owner still conspicuously stuck to the frame: "PROPERTY OF MOODY'S INVESTORS SERVICE LTD, LONDON".

This actually shouldn't have been so much of a surprise. Of all the computers I have parted with in my life, most were still in working condition and those that were broken looked far from damaged on the outside, while most of their operational parts also remained undamaged. With corporate and university devices, which seem to dominate the equipment on offer in the Computer Village, this is even more often the case (see also Forge, 2007). In retrospect, the idea that we were going to find piles of crushed and smashed devices coming off a ship from Europe was somewhat naive[2].

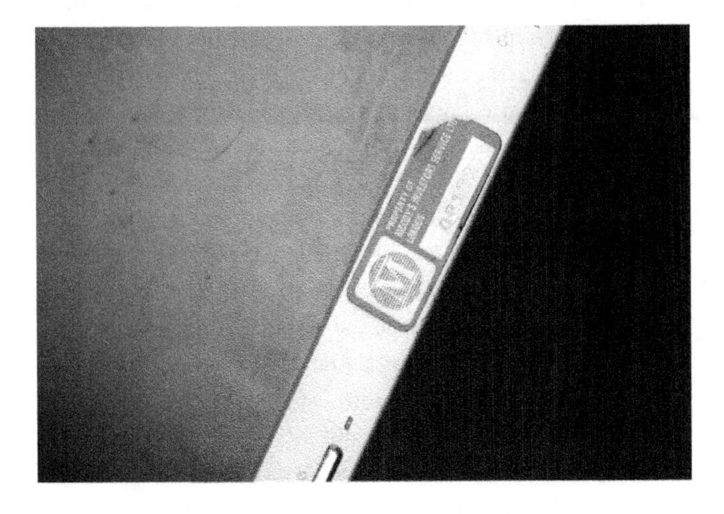

Fig.1: Used computer monitor with corporate sticker in Lagos, Nigeria,

Unlike products from the early days of the production logic of planned obsolescence (London, 1932), such as light bulbs and nylon stockings, contemporary consumer technologies – mobile phones, computers, printers – are often removed from people's everyday lives in the Global North before they have reached a state in which they show clear signs of physical decay.

Whereas stockings and light bulbs actually look broken when they are disposed of, inkjet printers are often equipped with a chip that causes the device to 'break' when a certain page

---

[2]In fact, Nigeria banned the import of non-functional used electronic and electronic devices in 2011. Although the effectiveness of the legislation is contested, it has likely also played some role in the decline in imports of proper e-waste – rather than functioning second-hand devices – into the country.

count has been reached (Katz, 2011). Similarly, Operating System and other software updates frequently necessitate users to replace hardware while it is still fully functional in a material sense. This rapid replacement of electronic devices in post-industrial consumer culture also means that users increasingly experience digital technology as perpetually new, and primarily associate these devices with notions of connectivity, disembodiment and 'progress'.

The manufacturing logic of consumer products appears to have shifted from a mode of 'analog planned obsolescence', where products become obsolete in such a way that the user is confronted with their material transformation into something that appears 'broken', toward a strategy of 'digitized planned obsolescence', where the consumer experience of the artefact is sanitized from the moment of acquisition until its disposal and an engagement with the 'dirtiness' of perceptible material decay is precluded. This latter form of planned obsolescence is specific to digital devices, as it depends on automated processes that remain hidden from the user's perception (such as the page-counting chip in the printer, or system operating retardation effected by software updates) inside what Bruno Latour (1999) has called the "black box" of electronic devices.

Mary Douglas's (2002 [1966]) writing on the function of waste and dirt in the organization of societies offers insights into the possible broader significance of this sanitized experience of consumer technologies. Douglas suggests that relations between subjects and objects in society are organized as a symbolic structure in which the concept of dirt plays a central role. The notion of defilement can act as a way to discipline behaviour in areas where human control and enforcement of regulations are difficult. Defining dirt as "matter out of place", Douglas identifies a progression of waste from a stage of "dirt" to what she calls "common rubbish". In the dirt stage, rejected objects still have a degree of identity; they are perceived in relation to elements of the cultural structure but they have been rejected in order to secure this order. After a process of "pulverizing, dissolving and rotting", which removes all identity from the rejected matter, it becomes "common rubbish" and as such no longer poses a "threat to good order" (Douglas, p. 197).

In the transition of contemporary consumer technologies from their useful lifespan to their status as common rubbish, in advanced consumer societies the dirt stage seems to be less and less conspicuous in the user's experience and in many cases is virtually absent from it. The product is removed from everyday life before it shows noteworthy signs of decay, and the transition process to the dirt stage is literally migrated to geographical locations in the Global South, which remain largely hidden from view. In the context of Douglas's understanding of culture as a symbolic structure, the consumer experience of technological devices increasingly functions in terms of what I propose to call a "symbolic order of technological progress", which precludes an engagement with the bare materiality of consumer technologies as dirt and contributes to their construction as mere signifiers for immaterial and abstract notions such as social connectivity, wellbeing and innovation.

Central to the logic of this symbolic order of technological progress is the avoidance and erasure of signs of material ageing and decay, which establishes a sanitized experience for users of digital devices. This absence of perceptible material decay suggests that technological innovation and planned obsolescence can continue ad infinitum and bear no relation to impending ecological catastrophe and exploitative labor conditions. Such a drive for the elimination of signs of decay can also be found elsewhere in consumer culture, most notably in contemporary attitudes to human bodies.

Building on Julia Kristeva's (1982) writing on the abject, cultural theorist Deborah Caslav Covino (2004) draws attention to the obsession with obtaining a "clean and proper" body in contemporary consumer culture. Particularly in advertising for plastic surgery, but also in the promotion of anti-ageing cosmetics and fitness regimes, body parts represented with characteristics of aging-related physical decay (wrinkled faces, sagging breasts, belly fat) are considered abject: that which is positioned outside the symbolic order and is rejected by social reason. Advertisements and other sources in popular culture suggest that through the consumption of surgical interventions, fitness programs and cosmetics, these abject elements can be eliminated to obtain a proper, sanitized body.

With this dynamic in mind, I would like to consider the final destination of obsolete electronic devices. After they have been discarded by their first users, most of whom reside in the Global North, a large number of these devices end up on markets in the Global South, including China, India and Sub-Saharan Africa. In places like the Computer Village in Lagos, the devices are usually resold as secondhand products, sometimes after being repaired but often in exactly the state they arrived in, albeit cleaned and repackaged. Eventually they will end up on a dump site though, oftentimes somewhere in the vicinity of the secondhand marketplace where the devices were resold: functional obsolete devices exported from Europe and elsewhere are not shipped back after they finally break down properly. These final destinations in the Global South are the sites where the process of perceptible material decay takes place properly: Dysfunctional devices are slaughtered for recyclable parts, and the remains are left behind in pieces. This is the process Jelili Atiku and I finally got to observe – and take part in – on a journey a year after our exploration of the

Computer Village, when we visited the dumping and recycling sites next to Alaba Market on the outskirts of Lagos.

If we consider this decaying e-waste in relation to the logic of digitized planned obsolescence, the notion of the abject also seems appropriate: these artefacts are positioned outside the symbolic order of consumption and are rejected by social reason. Furthermore, in accordance with Kristeva's concept, a confrontation with this decaying e-waste threatens a breakdown of meaning of consumer paradigms in which engagement with electronic commodities is disconnected from its material consequences.

There are some differences between the notion of the abject in the world of consumer technology and the "clean and proper" body culture discussed by Covino: whereas in the former the abject is hidden right from the start to facilitate the illusion that electronic devices are disconnected from their material context, the latter model is based on a continuous heightening of experiences of the abject within consumer culture in order to stimulate consumption patterns focused on expelling these abject elements. However, in both cases a process of sanitization through the elimination of abject elements is central to the stimulation and acceleration of consumption. 'Body beautiful' incentives and digitized planned obsolescence both contribute to a cult of timeless, enduring newness and youth that is driven by uninhibited consumption.

3. Hi-Tech Everything:

A report from the heart of techno-consumerism

In September 2017, I represented Philips at Europe's biggest consumer electronics fair, IFA, in Berlin. Curious to find out how the dreams of a smooth, technologized world are constructed and disseminated from the inside, I applied for a job with a PR agency that represents Philips Consumer Lifestyle at trade fairs worldwide. After attending an interview and submitting various kinds of photo and video footage I was hired to promote the latest innovations in so-called 'Male Grooming'.

Restrictions

Agreement, dated 18 August 2017

Between
1. XXX (the 'Client');
Dani Ploeger (the 'Freelancer')
The Freelancer has agreed to make their services available to the Client on the terms and conditions set out in this Agreement.
(...)
10: Confidential Information
(...)
10.1.9 All notes, memoranda, records, documents and other works (whether recorded on paper, computer memory or discs or

otherwise) made by the Freelancer in the course of their duties or relating to the activities of the Client and any copies thereof or other records prepared from such notes, memoranda, records, documents and other works or information contained therein shall be and remain the property of the Client and together with any Confidential Information in the Freelancer's possession, custody or control shall be promptly handed over by the Freelancer to the Client on the date on which this Agreement is terminated and at any stage during the term of this Agreement when requested by the Client.

10.1.10 The provisions of this clause 10 shall survive the termination or expiry of this Agreement for whatever reason.

Fig. 2: The author promoting a Philips S9000 shaver at IFA 2017 in Berlin (video still).

Notes on a public document made by a third party

At the trade fair, I promoted the Philips 9000 Series shaver with V-Track Precision Blades PRO, 8-directional ContourDetect technology and Aquatec, among other innovations in so-called male grooming technology. Dressed in a neatly ironed shirt with the company logo on the front and the text "How can I make your life better?" printed on the back, I spent seven days addressing thousands of consumers and retailers through a little portable loudspeaker. My contract, an excerpt of which is included above, does not allow me to write about the notes and documentations I made during my work at the Philips stand. However, an anonymous third party made a video of me and posted this online[3]. The following are some thoughts in response to this video, which do not form part of the 'notes, memoranda, records, documents and other works' I ('the Freelancer') made as part of my duties for 'the Client' (see paragraph 10.1.9).

Like the shaving devices, most products on the Philips stand are domestic technologies focused on health and well-being: robotic vacuum cleaners and air purifiers (3000 series with 'VitaShield IPS' and 'AeraSense technology'); low-fat air cookers (with 'unique Rapidair Technology'); data-collecting baby care products ('AVENT uGrow baby development tracker'); electric toothbrushes that monitor your brushing technique through a phone app (Sonicare FlexCare Platinum Connected). These domestic technologies, which in various ways are related and connected to their users' bodies, are promoted under the overarching slogan that appears across the stand: 'There's Always a Way to Make Life Better'.

The connection between the body and a desire for improvement that is established through this marketing effort is reminiscent of Zygmunt Bauman's (2000) discussion of the transition from a culture organized around a notion of 'health', towards an engagement with bodies that is shaped around the concept of 'fitness'. Health has traditionally operated as a

[3] http://bit.ly/Ploeger05

normative principle, marking a boundary between a state of normality and abnormality (disease), which can be described and often measured to a degree of precision. Being 'in good health' has been connected to a sense of the body being in an adequate condition to meet certain socially determined requirements, such as acting out a profession. On the contrary, the concept of 'fitness' is connected to subjective experience and not tied to a measurable, finite state. There is no end to fitness, you can always be 'fitter'. Accordingly, Bauman points out that 'Life organized around the pursuit of fitness promises a lot of victorious skirmishes, but never the final triumph' (2000: 78).

Bauman considers this to be a symptom of what he calls 'liquid modernity', a situation in which the organizational and power structures of modern society have become unstable in a way that promotes insecurity and anxiety tied to ever-increasing consumption. Thus, the attachment of the slogan 'There's Always a Way to Make Life Better' to body-related technologies conveniently uses liquid-modern consumers' anxieties around their ever-elusive fitness goals to incentivize them to continuously replace their domestic consumer technologies with the latest, improved version.

Another point of interest in the way in which domestic technologies are promoted here relates to Russell Davies's (2009) concept of the postdigital. Davies describes the postdigital as a condition where society has been saturated with digital technologies and where they have been integrated into the everyday to such an extent that their presence starts to lose prominence in people's perception. At the same time, concepts developed as part of digital on-screen and online formats start to be implemented in other aspects of life in the physical world.

The fact that the toothbrushes, vacuum cleaners and baby care products mentioned above have now all entered the domain of advanced digital technologies might be an indication that we are indeed at the dawn of a postdigital era. At the same time, the way in which the shaving machines in particular are promoted suggests that the transfer of principles originating in the realm of digital technology goes beyond merely "stuff that digital technologies have catalysed online and on screens" (2009: n.p).

The terminology used in relation to the shavers' technical aspects (V-Track Precision Blade PRO, 8-directional ContourDetect technology), as well as the model designations of the devices ('7000 Series', '9000 Series'), resemble the jargon that was in previous eras reserved for high-tech innovations in the specialist domain of scientific, military and personal computing technology. Note the formal similarity with names used for high-end computer components, such as 'ASUS H170 PRO Gaming ATX Motherboard' or 'Intel Core 2 Quad Processor Q6600'. The spill-over of this jargon into the realm of shaving devices means that this product, which previously might have been considered as a rather unspectacular, long-life home appliance, is now increasingly likely to be perceived as belonging to the realm of semi-disposable high-tech items that operates according to a logic of rapid innovation.

Thus, it appears that a postdigital condition will affect the broader field of commodity marketing as well. While the widespread integration of digital technologies into everyday life leads to a decline in users' conscious experience of their presence, this also facilitates the dissolution of clearly perceptible boundaries between digital commodities and devices that were traditionally considered as long-life products. At last, we might see the incorporation of *all* commodities into the paradigm of rapid product obsolescence and high-speed consumption.

Postscript

12.1 The Freelancer shall not, whether directly or indirectly, or whether on their own behalf or on behalf of any other person, firm or company, or as agent, director, partner, manager, employee, consultant or shareholder of or in any other person, firm or company:

(…)

12.1.2 at any time after the termination of this Agreement, in any way hold itself or himself out as engaged by, representing or acting for the Client.

4. Eerie Prostheses and Kinky Strap-Ons:

Mori's uncanny valley and ableist ideology

It's a spring evening in the early 2010s when I browse through the assortments of sex shops in London's Soho together with queer feminist porn star Zahra Stardust, in preparation for a collaborative artwork. Whilst we never end up completing the work, we envisage a performance in which she is going to anally penetrate me from the other side of the world with a telematic fucking machine. In the shops, we look at various prosthetic limbs: rubber arms with clenched fists, penis extensions and strap-on dildos, many of them in pink and brown skin tones and with vein textures in relief for added realism. Looking at these toys as prosthetics makes me think of roboticist Masahiro Mori's uncanny valley hypothesis.

In his short article 'The Uncanny Valley' (2005 [1970]), Mori reflects on the difference between the way industrial robots – developed with a focus on functionality – and humanoid robots – designed with an interest in human likeness – are experienced. Whereas industrial robots bear little resemblance to human bodies and usually do not evoke a sense of 'familiarity', toy robots with limbs and facial traits that resemble the composition of a human body are more frequently experienced as familiar.

However, Mori argues that the experience of familiarity with robots and prosthetic limbs does not increase linearly according to their degree of human likeness. If likeness is further increased beyond the example of the humanoid toy robot, as is the case in certain cosmetic artificial limbs and humanoid robots such as Repliee Q1 and Q2[4], the robot or prosthesis is experienced as uncanny[5].

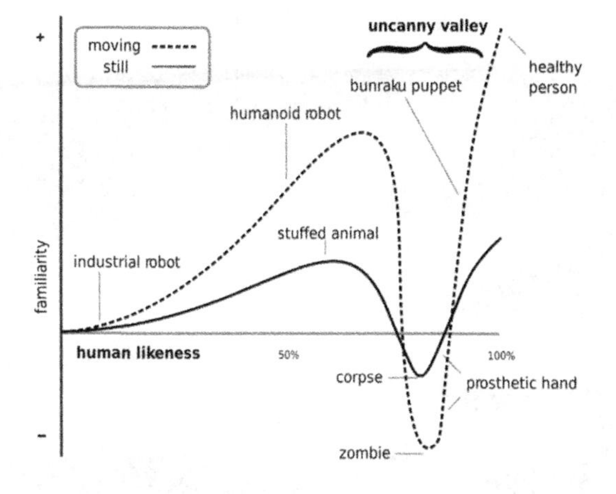

Fig. 3: Mori's uncanny valley (Mori 2005 [1970])

Mori suggests that whereas prosthetic limbs built with a focus on functionality evoke a sense of 'sympathy', a prosthetic hand that shows great likeness in appearance with a 'real' hand is uncanny. The uncanny valley then, is apparent in a graphic representation of different degrees of human likeness in robotics and prosthetics design, set out against the degree of

---

[4]Repliee Q1 and Repliee Q2 are so called Actroids, sophisticated humanoid robots that have been developed at Osaka University, Japan, in collaboration with Kokoro Company Ltd. since 2002.

[5]Mori does not define the term 'uncanny' in his paper. Throughout this text, I use the term in accordance with Oxford Dictionary of English's (2010) broad definition as "strange or mysterious, especially in an unsettling way".

familiarity they are expected to evoke (figure 3). Applying this graph as a guideline in technology design, Mori concludes that designers should avoid the uncanny valley by means of developing technologies with a "safe familiarity by a nonhumanlike design" (2005 [1970], n.p.).

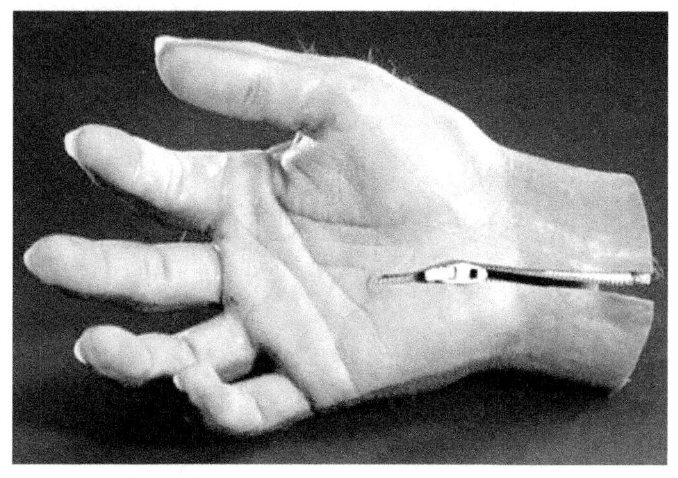

Fig.4: Latex prosthetic hand, 1970s. Smithsonian National Museum of American History.

Mori's hypothesis seems convincing when we consider realistic prosthetic hands (figure 4) and recent humanoid robots. However, the skin-coloured and veined strap-on dildos we find in Soho don't feel any more uncanny to me than their more abstractly shaped, pink and purple counterparts. Regardless of the detail with which manufacturers simulate veins and skin pigmentation, these artefacts seems far less likely to be associated with the uncanny than the prosthetic hands Mori alludes to in his text. The realistic aspects of the example here (figure 5) afford a sensation of kinkiness rather than of uncanniness. Does this suggest that the uncanny valley hypothesis is false or inaccurate?

In a recent study, psychologists Burleigh, Schoenherr and Lacroix (2013) presented experimental subjects with a range of computer-generated representations of human faces. They performed two experiments. In the first experiment, subjects were presented with a continuum of facial representations with increased geometric realism and prototypicality (i.e. adherence of the face's shape to common expectations of human appearance). Participants were asked to rate the degree of human likeness, as well as the degree of 'eeriness'[6] they experienced for each of the generated facial representations. The outcome of this experiment showed that human likeness and perceived degrees of eeriness are linearly correlated; the more human-like the image, the less eerie (or more 'familiar') it was experienced to be. In other words, the experiment suggested that Mori's uncanny valley hypothesis does not hold true.

In order to explain how earlier empirical research outcomes by other psychologists could have been interpreted in support of the uncanny valley hypothesis (MacDorman & Ishiguro, 2006; Mitchell et al., 2011; Saygin et al., 2011; Seyama & Nagayama, 2007) Burleigh et al. conducted a second experiment. For this experiment two continua of human likeness were generated. The first was limited to what the researchers describe as the "ontological category of humans" (2013: 761), whilst atypical human features were introduced along the continuum ('unnatural' skin colour and an enlarged eye). The second continuum merged human and non-human categories and featured hybrids of human faces with those of a goat-like creature. Responses to the first continuum were comparable to those in the first experiment, i.e. there was a linear correlation between perceived degrees of human-likeness and eeriness.

---

[6]Burleigh et al. use the more colloquial term 'eerie' to describe the uncanny in their experimental questions.

However, the continuum introducing non-human features showed a heightened level of eeriness and lower level of pleasantness around the mid-point of human likeness. Based on these findings, Burleigh et al. suggest that the uncanny valley is not directly related to the degree of human likeness. Instead, the valley seems to occur where stimuli located at a mid-point between two ontological categories elicit a negative affect due to ambiguous and conflicting interpretations; uncanniness results from a 'category conflict' where the subject is unsure whether what is perceived is human or non-human.

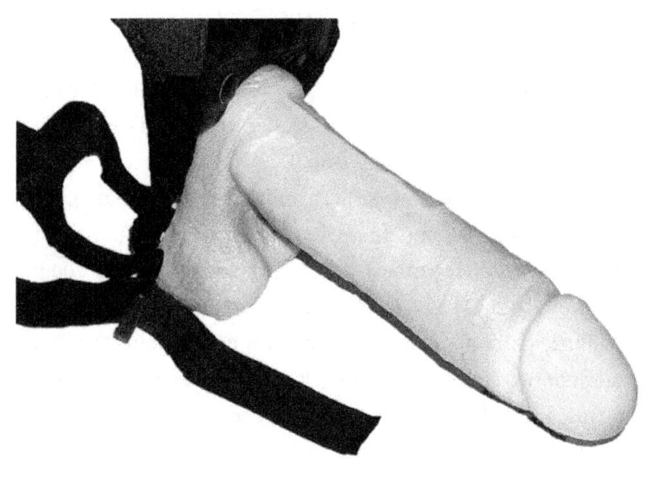

Fig. 5: Strap-on dildo

Now, let's return to the strap-on dildo and the realistic-looking prosthetic hand. Why is the prosthetic hand more likely to evoke a sense of uncanniness than the realistic looking dildo? Following Burleigh et al., the hand may be more likely to evoke a 'category conflict' because the beholder is unsure whether the hand is an actual human body part, or a non-human substitute for a missing 'real' hand. The strap-on does not pose this difficulty; it is usually perceived as an addition to an 'able', whole body. Regardless of the degree of human likeness of its appearance, there is no doubt about the non-humanness (or 'unnaturalness') of the dildo, and accordingly it is unlikely to evoke an uncanny experience.

As mentioned above, in the conclusion of his paper, Mori calls for designers to steer away from high degrees of human likeness in their artefacts, in order to avoid the uncanny. Pointing to their finding that the uncanny valley is not a result of a higher human likeness of features, but of ontological category conflicts, Burleigh et al. recommend that there is no reason for digital designers not to aim for high degrees of human likeness in terms of graphical texture resolution and polygon count of computer-generated models. However, they recommend avoidance of the combination of human and non-human features. This may sound like a sensible suggestion if we accept that categories of the uncanny are stable and a confrontation with something we experience as uncanny is by definition negative. However, on paying closer attention to the concept of the uncanny this position is less self-evident: if we consider psychologist Ernst Jentsch's account of the concept it becomes apparent that the uncanny should not necessarily be understood in ontological terms.

In what is generally regarded as the first publication on the concept of the uncanny – predating Freud's writing on the topic – Jentsch (1997 [1906]) argues that experiences of the uncanny are triggered by an uncertainty concerning the animatedness of an object. Referring to a human skeleton, he suggests that uncanny objects evoke "thoughts of latent animatedness" (1997: 15). Thus, in a somewhat similar vein to Mori, Jentsch attributes the experience of the uncanny to the closeness in appearance of the skeleton to aspects of a living human body; the skeleton looks as if it *could* be animate. Thus, instead of resulting from a certain relation to a stable ontological realm, when we follow Jentsch's perspective the uncanny should be seen in the context of a symbolic system, which needs to be *believed* to be true or realistic by the subject.

If a sense of the uncanny is an affect that is dependent on a subject's belief in a certain constitution of the 'real' world, and may disappear either through habit or an associative working through, the 'category conflict' Burleigh et al. refer to should not necessarily be considered of ontological nature. Indeed, when approached from the perspective of theories in posthumanism (e.g. Hayles 1999; Wolfe 2010), the categorization of what is 'human' is neither universal nor stable.

The difference between the prosthetic hand and the strap-on dildo is that the former is perceived as a substitute for something believed to be *missing* from an 'incomplete' human body, whilst the latter is more likely to be regarded as a kinky addition to an 'able' body that is 'whole'. In other words, the uncanniness of the realistic looking prosthetic hand lies in its function as an indexical signifier for bodies that are considered incomplete.

Cary Wolfe points out that despite its intention to promote the treatment of people with disabilities with equality and respect, humanism reproduces "the very kind of normative subjectivity – a specific concept of the human – that grounds discrimination against [...] the disabled in the first place" (2010: xvii). The fact that an essentialist, unchangeable concept of the human body is at the very centre of humanist thought means that bodies that do not correspond to this model are by definition classified as different and unequal. In this context, the uncanniness of the prosthetic hand can be seen as a manifestation of an essentialist and ableist idea of 'the' human body, which is narrowly defined in terms of a stable set of characteristics (Wolbring 2008).

Following Jentsch and posthumanist critiques of essentialist concepts of the human body, a repetitious experience of – or conscious engagement with – the uncanniness of the prosthetic hand would facilitate the integration of the prosthetic hand in the symbolic system of our concept of the human body. It would no longer act as a signifier for a lack of 'normal' humanness in the context of a fixed idea of a 'whole' body with two hands. Thus, just as a body without a strap-on is not commonly perceived as incomplete or disabled (*pace* psychoanalytic suggestions of female bodies 'lacking' a penis), the 'un-uncannying' of the prosthetic hand could play a role in the acceptance of a body with one or no hands as a normal human body.

The advice of Mori and Burleigh et al. to designers to avoid the uncanny in their work may be sensible from the perspective of product sales optimization in the current market economy. However, this approach eventually reinforces the narrow set of cultural beliefs of what constitutes a 'whole', 'healthy' human body, which is central to the discrimination of bodies that fall outside this constructed category of the normal and the natural. Instead, wouldn't it be preferable for artists and designers to aim for the uncanny, and stubbornly forge 'category conflicts' where commonplace ideas of 'the' human body are merged with cultural categories of the abnormal, unnatural and unhealthy? Let us produce so many uncanny artefacts that the 'force of habit' will teach us to experience an infinite variety of bodies as normal. Not only will this transcend prejudice against disabled bodies, it would also be the dawn of a body culture where the norms for physical modifications are no longer determined by anthropocentric ideas of humanness as a confined concept. As such, it will contribute to a fluid and inclusive perspective on the place of humans in the world, beyond exploitative behaviour rooted in fantasies of human supremacy.

5. The Dirt Inside:

Computers and the performance of dust

One day in 1987 or 1988, my mother performs her usual domestic cleaning routine in our house in the Dutch countryside: every second day, the house is vacuumed and dusted. Normally, this work would remain largely unnoticed and my father would hardly acknowledge that the spotless and dust-free condition of the contents of his 'computer closet' are the result of a disciplined domestic work schedule. However, this is a special day. A minor crisis emerges after my father notices, after coming home from work, that one of the keys of his Sinclair ZX Spectrum+ has been sucked into the vacuum cleaner. My mother, who has never shown any interest in the device, and would never touch a computer for other purposes than cleaning, hadn't noticed. "Your mother is clueless about technology", my father explains.

In April 2015, I dismantle laptop and desktop computers at Vannex International Ltd., an electronic waste recycling factory in Hong Kong. Together with a group of artists, cultural theorists and scientists from Nigeria, Hong Kong and across Europe, I followed discarded electronic devices from the heart of consumer culture to dumps and recycling factories in Nigeria and Hong Kong. As part of a practice-based methodology, we participated in recycling work on a dump in Lagos and in this recycling factory in Hong Kong. Initially, I wondered why I was asked to wear a dust mask in the factory in Hong Kong: the computers piled up on the work floor appeared to be clean and pretty much all of them were in undamaged condition. However, once I opened their cases to remove their components, clouds of dust emerged.

On 9 September 2015, I wake up next to princess S. Her hair is no longer pink, after a summer in the desert kingdom. We fucked amidst a collection of various state-of-the-art Apple computers, financed by the royal family and its pharmaceutical enterprises. My ripped skin is under her nails, condoms with blood on them lie on the floor and bed. The computers are dusty, the keyboards covered in grime. Her art school professors rejected her digital work: it lacks focus, they say.

In advertisements, computers are often associated with cleanliness. In some cases this is explicit, as in a Lenovo commercial[7] that presents their devices as a magical means to clean-up chaotic offices. Likewise, an Apple commercial from the mid-2000s[8] shows a computer that creates dust and rubble but stays conspicuously clean itself.

However, the three anecdotes above show that the everyday life of these machines may not be as effortlessly clean as their PR image suggests: the celebrated cleanliness is achieved through an – often unnoticed – meticulous cleaning regime; cleanliness tends to only concern the outer appearance of the devices; and computers may also exist in a liminal realm between royal power, capital decadence, and sexual transgression.

Sociologists Gary Alan Fine and Tim Hallett (2003) suggest that understandings of larger social systems can be gained

---

[7]http://bit.ly/Ploeger08

[8]http://bit.ly/Ploeger09

through an examination of the smallest conditions of life, by practising what they call 'sociological miniaturism.' Their interest is in the small material par excellence: dust. What if we try to understand consumer computing not through its commonplace, 'official' narrative of the clean and the pristine, but through its inevitable and persistent bond with dust?

Dust and morality

Dust as the smallest perceptible particle has long been associated with notions of moral purity and control, especially in the domestic sphere. Building on her concept of dirt as matter out of place, anthropologist Mary Douglas (2002 [1966]) suggests that the presence of dust indicates a loss of control over the living environment, ultimately posing a threat to our sense of moral order. This is aptly illustrated by the following examples presented by Fine and Hallett:

> A clean, fresh and well-ordered house exercises over its inmates a moral, no less than a physical influence and has a direct tendency to make the members of the family sober, peaceable and considerate of the feelings and happiness of each other; whereas a filthy, squalid, unwholesome dwelling, in which none of the decencies common to society are or can be observed, tends directly to make every dweller regardless of the feelings and happiness of each other, selfish and sensual.
>
> (English reformer Southwood Smith in Fine and Hallet, 2003)

> [Cleanliness is the] "parent of virtues." [It is] an emblem, if not a characteristic, of purity of thought and propriety of conduct. It seems as if it could not be associated with vi-

> cious pursuits; so rarely, in the habitually profligate character, are the active and wholesome habits of cleanliness perceptible.
>
> (*Encyclopedia of Domestic Economy* (1844) in Fine and Hallet, 2003)

Furthermore, contemporary research by Dunifon, Duncan, and Brooks-Gunn (2001) suggests a positive correlation between the cleanliness of a child's home, the number of years they spend in education, and the size of their income. Could this be a confirmation of Max Weber's (1958 [1905]) thesis on capitalism and the protestant work ethic applied to the domestic sphere? Whilst these research projects may be somewhat questionable in terms of the bold conclusions they draw on the basis of somewhat shaky statistical interpretations, they do suggest that there is at least a strong *perceived* connection between a dust-free environment on the one hand, and pristine morality, economic and intellectual progress on the other.

In the context of domestic and office computers, such connotations support and propagate a utopian, technologically determinist perspective. The Lenovo and Apple commercials I referred to above both suggest a concept of technology as the driving force for change in the (human) environment. A representation of the devices as ever shiny and dust-free connects this deterministic force with a sense of the morally pure: technological change is not only inevitable, it is intrinsically good. In turn, these representations may become simulacra that inspire an obsessive domestic (or office) cleaning regime. In *The Second Sex* (1953 [1949]), Simone de Beauvoir suggests that such dedication to dusting and cleaning prevalent among housewives often facilitates a flight from themselves, which "may often have a sexual tinge… Love of the flesh and its animality is conducive to toleration of human odour, dirt, and even vermin." (p. 452)

Dust and computing performance

Dust on integrated circuits prevents a computer's cooling system from operating adequately. The increased temperature of the circuits will cause the system's performance speed to decrease and ultimately leads to system crashes. Conversely, the cooling fan inside the computer animates the dust and stirs it up when it is switched on. In other words, there is a mutual interaction between the dust and the computer, which is directly connected to the device's operation.

If we consider that household dust consists of 70-90% human skin flakes (Clark and Cox, 1973) and follow Donna Haraway's definition of a cyborg as "a cybernetic organism, a hybrid of machine and organism" (1985), this leads to an unexpected perspective. Looking at it like this, could an ordinary computer actually be considered a cyborg? This would be a cyborg that neither techno-utopians like Stelarc and Kevin Warwick, nor critical thinkers such as Haraway and Catherine Hayles have considered: instead of human bodies with agency that are extended with state-of-the-art technology, we are now considering machines extended with lifeless human body parts.

In his book *On Garbage* (2005), John Scanlan proposes a concept of garbage that goes beyond a literal concern with physical material. He discusses garbage as a metaphor, but also – and more relevant to my interest here – the notion of the 'garbage of knowledge'. Discussing Kant's *Critique of Pure Reason*, Scanlan draws attention to the vast amount of Kant's notes which have been preserved but were not included in the book. The production of knowledge is a process of creating garbage: discarding those thoughts and ideas that turn out not to fit in or are not consistent with the paradigm.

I propose to extend Scanlan's concept of the 'garbage of knowledge' to the realm of concepts and imagination. From this perspective, the dusty computer can also be seen as the garbage of the concept 'cyborg': that which is consistent with the concept, but is rejected or has been discarded from our imagination.

The dirty computer as activist creature

When I recall the bedroom of Princess S. mentioned in the beginning of this text with these deliberations in mind, the wild domestic environment she invited me into gains another dimension of meaning. Rather than interpreting her dirty devices as a sign of bad hygiene, I want to embrace them as companion species of sorts, garbage cyborgs that contribute to an embracing of the "love of the flesh and its animality" through its connection to "human dirt", to use Simone de Beauvoir's words. Opposing the cult of sterile, dust-free computing and its implicit propagation of efficient, infinite technological innovation as morally desirable, S. and her hybrid creatures offer a vision of a techno-culture beyond alienation, where digital devices become an integrated part of our lifeworld, a technologized ecosystem where our computers slow down and crash amidst the dust and fluids of our bodies in an orgy of inefficiency and dirty freedom.

6. Orodha:

The ultimate fetish commodity and its reversal

The Swahili word orodha can be translated as 'junk' or 'waste'. In Kenya, it is mostly used to designate a particular domain within the trade in second-hand products. Orodha items have usually been intercepted at some point on the way from their existence as everyday use-objects to becoming refuse on the garbage dump. They are often broken or incomplete and in the case of electronic devices, the salesperson as a rule won't be able to tell you whether something works or not. The term orodha is used to designate these objects, but also their traders and the places where the business takes place.

On the orodha market, one can find almost any imaginable everyday object, ranging from television remote controls and kitchen devices to toys and plumbing articles. For many things the use-possibilities are clear. An old remote can serve as a replacement, a broken kitchen blender can be harvested for spare parts, and plastic dolls can be readily used by kids to play with, even if some limbs are missing. However, there are also orodha items for which the possible use-value is not that straightforward.

During a visit of the orodha market in Kariobangi North, a suburb of Nairobi, together with artists Greenman Muleh Mbillo and Joan Otieno, I find the remains of a slide projector. It can no longer be restored to its original state, because too many parts are missing. Besides, there are no slides on offer in the stall where it is sold and neither the salesman nor any bystanders are aware of anybody who might have any.

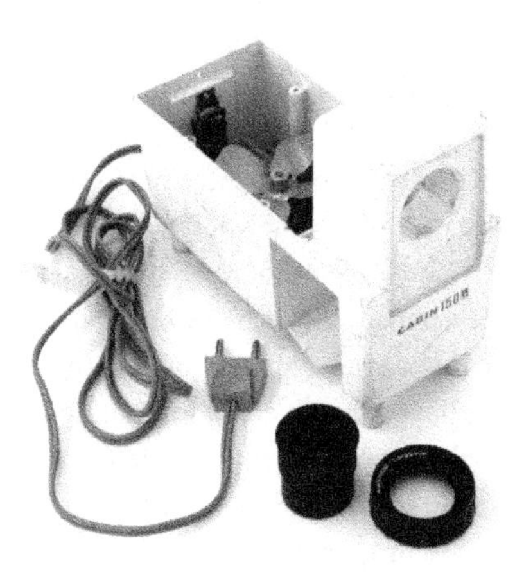

Fig. 6: Remains of slide projector acquired in Nairobi, 2019

Of the device's electrical components only the cooling fan and the mains cable remain. The fan still works and its metal blades make it look like a robust component by today's standards. However, it can hardly be used as a spare part in contemporary electronic devices. Their fans usually operate at 12 Volts, while the one in the orodha projector is powered with 220-240 Volts, straight from the power grid; the transformer is integrated in the motor. This also means that the object is unsuitable as a kids' toy: they'd likely electrocute themselves. When queried, the salesman tells me that "it's a projector", but when I ask whether he thinks it can be repaired he just laughs at me. He then takes a projector lens from his pocket, which he puts into the hole in front of the object. It doesn't really fit and clearly belongs to another device, but that doesn't seem to be relevant.

In the marketplaces I frequent in the Global North, this object would most probably have been discarded long ago. With no apparent remaining possibilities for use, it would be ejected into the realm of what anthropologist Mary Douglas calls "the heaps of common rubbish", (1966) the final destination of discarded use-objects where they lose their individual identities and become part of a non-descript mass of materiality. Instead, as orodha, the object seems to have a liminal identity that hovers somewhere between this state of common rubbish and a commodity that has a designated place in market exchange. While there seems to be little – if any – possibility that a use-value will be identified by either seller or potential buyer, the object is still put on display as a commodity-for-sale in a prominent place in front of the stall. Moreover, although restoration is acknowledged by the seller to be impossible, the original identity of the object as a slide projector is maintained by providing potential buyers with a projection lens, albeit one that does not fit.

Considered from this perspective, the object becomes somewhat mysterious. How are we to understand this apparently useless thing's status as a commodity that is traded on a market? The Oxford English Dictionary defines a commodity as something "that can be bought and sold" but also as a "useful or valuable thing" (Oxford Dictionary of English, 2010). The former does apply, but the usefulness of this object remains unclear. Hence, if we are to consider it as 'valuable' this would most likely only be in terms of what Karl Marx (1974 [1867]) calls 'exchange-value', the value in market trade exchange, usually expressed in money. There doesn't seem to be any 'use-value' in Marx's terms. As such, we could say that it's the commodity fetish object par excellence. For Marx, 'commodity fetishism' is a perception of the relations of production and trade exclusively in terms of monetary exchange, instead of a social process in which people's interactions are central. Thus, commodity fetishism promotes a process of alienation where social interaction loses relevance in our interaction with things. In case of the orodha technology we are considering here, not only the realm of production, but even the social potential of practical use seems to be excluded from its perceived value on the market.

But is this really all there is to it? The first objection to this reductive interpretation would be that, due to its apparent lack of use-value, the process of trading this object is actually highly focused on the social interaction between seller and buyer. Since the seller seems to have little idea what the object might be used for, his estimation of an appropriate price appears to be almost entirely based on his judgement of the potential buyer through his interaction with them. What does their behaviour disclose about the value they might attach to this object? What does the way they dress – or the colour of their skin – suggest about how much they would be able and willing to pay? The commodity itself appears to move to the background in this process, so rather than becoming a trope of alienation it actually acts as a catalyst for social interaction.

It may be more fruitful to take an altogether different viewpoint though. Instead of looking at this object as 'lacking' use-value and therefore designate it as a merely symbolic – and potentially alienating – entity, we may also consider the apparent absence of use-value as a material space of opportunities. The object is still connected to its original purpose and the allure of the paradigm of consumer technology it originates from. This is emphasized by the lens that is provided with it. It doesn't fit and there is no hope of restoring the object to a working projector, so – as mentioned above – the role of the lens should probably be understood primarily as a way to frame the object in terms of its origin as a consumer technology. As such, the new owner becomes part of the world of high-tech consumerism. However, this seemingly symbolic process has a material dimension. The buyer does not merely become a consumer of a pre-manufactured technology. The trading process implies that they need to do *something* with this thing, but what exactly, remains unspecified.

The object becomes an invitation to embrace technological remains and indulge in their – at first sight – apparent uselessness. Stripped of any possibility of regaining functionality in terms of its original identity as a typical Euro-American consumer technology device – a domestic slide projector – the object becomes a material stimulus to imagine and create a technology that goes beyond the commonplace logic of everyday techno-consumerism. Instead of responding to a set of ingrained affordances of a readily functioning device, the clearest way to engage with this object is through experimentation and play. What could we make with it and what could it do? What could everyday technology become, once it drifts off the path of self-replicating consumerism?

7. Frugal Phone / Material Medium

The X-Tigi S23 mobile phone is sold in every other phone shop in Kenya. Depending where you buy it, it costs anywhere between the equivalent of €16 and €30. With its large number buttons, low resolution colour LCD screen, flashlight and communication functions that are practically limited to voice calls and SMS messages it is reminiscent of the classic mobile handset models from the beginning of the century. But it's much bigger due to its huge battery, part of which conspicuously sticks out of the back of the device. It only needs to be charged once a month. Its square design with faux-metal nail and screw patterns on the sides evokes a sense of robustness.

At first sight, the X-Tigi seems a typical example of what is often called 'frugal innovation' in economics (Zeschky et al., 2011). Responding to the everyday conditions of consumers with limited economic means and in challenging environments, this kind of innovation involves stripping down products to their essential features, rather than adding ever more technologically advanced functions. However, when we take a closer look at the S23, it becomes apparent that the differences between this device and state-of-the-art smartphones might not merely concern a reduction in functions and functionality.

The most prominent feature of a smartphone is doubtless its screen, which nowadays covers almost the entire front of a device. Its other physical features – a few, mostly inconspicuous buttons, and openings on the front and back for the cameras and microphone – are overshadowed by the screen, which primarily acts as a gateway to a realm of networked data, which is detached from the direct material surroundings of the device or – in the case of taking pictures – seeks to mediate these surroundings for networked dissemination. In-ear headphones, which are the only accessory provided with most smartphones, let the user shift their engagement with their immediate environment further towards the informational realm of the network.

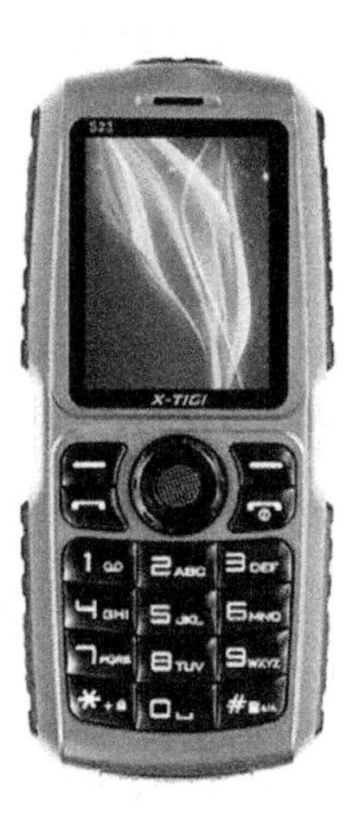

Fig. 7: X-Tigi phone

The X-Tigi seems to operate differently in several ways. The standard retail package of the device does contain earphones as well, but it also comes with an LED light bulb attached to a power cord, which can be connected to a USB port that features prominently on the back of the phone. It is marked 'OUT' with capital letters. This USB port can also be used to charge other – possibly smarter – devices in the direct vicinity of the X-Tigi. A large flashlight on the top of the phone can be controlled via on-screen menus, but in addition there is a sliding button on the side of the handset to control it directly, so there's no need to engage with the software interface. The common character of these features is that they seem to primarily revolve around direct interaction with the immediate environment of the device, either by illuminating it or by powering other devices that are physically present in the same space.

Another notable difference from most smartphones concerns the design of its casing. Whereas smartphones tend to have inconspicuous, smooth surfaces that ensure that user attention is solely focused on the screen, the X-Tigi has multiple features that seem to deliberately draw attention away from the screen towards the device's object-ness. The big number and control buttons cover more surface than the screen and the phone's plastic casing is marked by rugged shapes on the sides and fake screw reliefs on the back. Its robust appearance, which is heightened by these features, earned the device the nickname 'military phone'.

As such, the X-Tigi is not just a stripped-down version of a state-of-the-art mobile technology. In various ways, it appears fundamentally different from a contemporary smartphone. The latter draws the user into an experience of everyday technology as a realm of immaterial connectivity and endlessly floating information, a perspective that is coherent with consumption alienated from its material, environmental implications.

The user-experience of the X-Tigi phone, on the contrary, to a considerable extent revolves around direct connections with the materiality of its immediate surroundings and a design that emphasizes its own presence as a material object. It is a medium of materiality in the most literal sense. Thus, it may be a useful antidote to the ideology of smoothness and lightness propagated by the smartphone.

8. Positioning the Middle of Nowhere:

GPS technology and the desert

> The majority of units had been issued [...] an electronic device much like a pocket calculator which, using signals from satellites, could tell you your exact coordinates. However, we did not have such a system and our luck began to run out [...] The Iraqi desert is like the ocean.
>
> Lieutenant Colonel Joseph P. Gallagher, 1991

> One wondered from where the surrendering enemy had come. They carried no equipment, water, or food. Yet, they walked south in the barren, featureless desert with no enemy positions in sight.
>
> Captain H. R. McMaster, 1991

Global Positioning System (GPS) enables us to determine our precise geographic location anywhere in the world. It was first used on a large scale by the US military during the Gulf War in 1990-91 and proved decisive in the ability to navigate the desert landscapes of Kuwait and eastern Iraq, giving American and coalition forces a tactical advantage over the defending Iraqi military. After the war, the use of GPS in the civilian domain rapidly expanded. Nowadays, the omnipresence of mobile devices has made satellite-based navigation a standard component of everyday life. Moving through urban and rural environments is increasingly done with the assistance of Google Maps, while numerous other digital services make use of satellite positioning with users hardly noticing. This development has also led to a broader shift in awareness of space in everyday life. The *possibility* to determine your precise geographical coordinates at any time affects spatial awareness regardless of whether you are actively using the GPS function of your phone or not. As a consequence, the experience of being in a space that is beyond precise localization, the feeling of being 'in the middle of nowhere', has become increasingly rare.

In the West, the absence of an accurate sense of geographical location has particularly been associated with the ocean and the desert. In Christian traditions, the combination of perceived placelessness and inhospitality to human life have rendered both these environments ambiguous symbolic tropes. The ocean is the domain of apocalyptic violence, as well as an expression of the unruly, unbounded character of divine nature. The desert is the environment of torment and trial, but also the place where one encounters God (McGinn, 1994).

In Hollywood cinema, the desert has often acted as a trope for the supposed unreliability of Middle Eastern subjects. In movies like *Lawrence of Arabia* (1962) and *Raiders of the Lost Ark* (1981), the "exposed, barren land and the blazing sands" serve as a motif for the supposed 'hot' passion and uncensored emotions of Arabs, in other words: a "world out-of-control" (Shohat, 1997, p. 32). As such, the desert, as a metaphorical site of projection, has played a substantial role in the propagation in popular cultural of 'orientalism' as discussed by literary critic Edward Said (1978): the application of clichéd, generalizing analyses to cultures in Asia, North Africa and the Middle East that constructs the orient as a universal exotic place inhabited by archetypical subjects. Said argues that in this way, orientalism operates as a political instrument that ultimately serves to justify Western colonial domination.

Related to this cinematic use of the desert motif, the desert is not only perceived as an unruly site in terms of its physical inhospitality. It is also experienced as a challenge to sensory perception. In the desert, appearances are unstable. Mirages disorient your sense of distance, and - as art historian John Van Dyke reported after a journey through the North-American deserts in the early 20th century - you may perceive 'a pink air, a blue shadow, or a field of yellow grass' (Van Dyke, p. 110). Likewise, H.R. McMaster, a US Army officer during the Gulf War, describes surrendering enemy soldiers who appear to come from nowhere (McMaster, 1991). As such, the desert challenges the European, enlightened confidence in the ability to deduce singular truths from observation of one's environment. The desert is a place where your senses cannot be trusted as a source of definite, unchanging information.

Looking at the role of GPS equipment used by the US military in the Gulf War in this context, what might initially have seemed merely a pragmatic technological innovation gains broader significance. It is a 'pocket calculator' of sorts that compensates for the disorienting impressions of the senses in a 'barren, featureless' landscape. With satellite navigation, finding your way in the desert is no longer a matter of 'luck', as an officer responsible for POW transport during Desert Storm seems to suggest (Gallagher, 1991). As such, GPS re-asserts confidence in empirical assessment of the environment, albeit with a technological tool, while simultaneously subduing the orientalist Hollywood fantasy of the desert as an unpredictable 'world out-of-control.' While the eyes of Captain McMaster still confused him about the origin of the Arab enemy, his battalion's GPS-facilitated surprise attack neutralized their threatening nature and made them give up in defeat.

There is an additional point of connection between satellite navigation and the orientalist dimension of popular cultural imaginations of the desert. The supposedly objective perception of space that is enabled by the use of GPS-coordinates in fact embodies a particular way of looking at the world that is rooted in colonial history and has inherently militaristic characteristics. Navigation based on a bird's eye view perspective has been a predominantly European practice, closely tied to the use of cartography in colonial conquest. This way of looking at space remains secondary in many parts of the world, especially in the Global South. Furthermore, GPS technology's promotion of orientation based on target points, irrespective of their surroundings, is an approach with close ties to military strategy (Kaplan, 2006). In this way, GPS technology seems to fit in with some of the colonial implications of orientalist tropes in Hollywood cinema mentioned above.

The transformation of the desert from a biblical place of torment and potential encounters with God into a satellite-navigated grid of coordinates also forms a decisive step in another development. Philosopher Edward Casey describes a 'site' as the 'emptied out residuum of place and space' where 'place [is] dismantled into punctiform positions' (Casey, 1997, p. 186). Looking at cultural imaginaries and practices related to the desert in the United States, literature scholar John Beck (2001) argues that a perception of the desert as a 'site' in this sense has enabled its incorporation in the capitalist frame of a waste landscape, close to a non-place. As such, the desert is deemed appropriate for any activity that is not desirable in areas that are considered cultivated, be it industrial development, military endeavours or extraction of raw materials. In this way, GPS navigation unifies the US military's localized experience 'on the ground' of the Iraqi and Kuwaiti landscape with the long-standing approach of American foreign policy to the region. Ultimately, US interests in the Middle Eastern deserts are based on its conceptualization as a wasteland, speckled with profitable punctiform areas: oil wells.

The proliferation of GPS-based navigation in the civilian domain since the 1990s can be seen as a legacy of its prominence during the Gulf War and the media attention generated around it. In this context, the convenience in navigation it has brought to everyday life gains an additional, less comfortable, dimension. As American Studies scholar Caren Kaplan (2006) has suggested, GPS promotes a militarized vision of public space, which can be traced back to its origins as a warfare technology. When we navigate the streets with Google Maps, our focus on the moving blue dot on the screen fixates our attention on the 'target' of our movements, at the expense of observing and engaging with the features of the environment we are in. The peculiar difficulty of recognizing street names on most digital maps (the more you zoom in, the more they disappear it often seems) further promotes this tendency; rather than looking for and reading street signs in your surroundings, you are encouraged to keep staring at the map continuously.

Following from this, the landscape is transformed into a 'site' with 'punctiform positions' of interest, very much as in Casey's analysis, and similar to Beck's perspective on contemporary approaches to desert landscapes. In effect, everyday reliance on GPS navigation promotes an experiential desertification of the environment. In this new desert, the points of resource extraction that formed the essence of the Kuwaiti desert as 'site' – i.e. oil wells – have been replaced by points of consumption and capital exchange: shops, businesses and mass-culture entertainment (e.g. attraction parks) are marked most prominently on Google Maps. What might once have been experienced as a common space, accessible to and serving the interests of the entire community, is represented on the map as an area that primarily exists to accommodate the interests of capital accumulation.

Thus, everyday satellite-based navigation can be seen as a gateway to a next level version of what architect Rem Koolhaas (2002) has called 'junkspace'. For Koolhaas, junkspace describes a physical urban environment, the properties of which are shaped to entirely serve consumer culture. While GPS navigation doesn't transform physical space, it draws you into a virtual overlay of space that encourages you to experience it as a consumerist non-place in a way that intensifies the implications of Koolhaas' architectural critique. Maybe GPS does not mean the end for experiencing 'the middle of nowhere', but rather its substitution for a technological successor. In the new middle of nowhere you always know exactly where you are on the map, but every site is increasingly experienced like any other area of extraction and accumulation.

9. Sounds of Violence:

The affective tonality of high-tech warfare

Acts of war are usually accompanied by particular sounds: explosions, air raid warnings, aircraft engines, the whistling of projectiles flying by. At the same time, especially in the case of high-tech warfare methods, incidents of violence are frequently connected to imagined, largely fictional soundscapes by people who don't experience their actual occurrence in their everyday lives. Based on popular cultural representations (Carnahan 2010), public relations material (General Atomics 2012), and news media reports (Central Office of Information for Home Office 1975), sonic impressions of technologised warfare are propagated that do not match with their actual sounds. The representation of US military drones is a prime example of this. Whereas Hollywood movies and promotional material predominantly feature slick sounds of jet engines and robotic motors, in reality the General Atomics MQ-1 Predator and MQ-9 Reaper – currently the most frequently used US military drones – are propeller planes. Through the propagation of polished soundscapes, the sonic imaginaries of drone warfare support the visions of infallible technological efficiency and clean warfare that surround official narratives of military technology.

In July 2018, I travel to Karachi, Pakistan, to speak with people who have experienced US drone operations in the Federally Administered Tribal Areas in the north of the country. Together with Karachi-based artist and curator Mehreen Hashmi and artist Yasir Husain I travel to Sohrab Goth, a suburb about 10 kilometres from downtown Karachi that is predominantly inhabited by Pashtuns, many of whom have recently arrived from the north of the country. Although it is easy to encounter people from northern Pakistan in Karachi – an estimate of 15% of the city's population are Pashtuns, many of whom have arrived in recent years (The News 2011) – it is challenging to find somebody who is willing to speak openly about their experiences of war and terrorism. The Pakistani secret service, ISI, and the army have a reputation of detaining anybody who is deemed even slightly suspicious in terms of posing a risk to the country's interests. Furthermore, Taliban factions have been very active in Karachi until recently and at times controlled large parts of the suburbs. The fighters and their sympathizers are still there, but their whereabouts and activities are largely unknown, resulting in a continuous sense of unease about a possible upsurge in activity or targeted actions against individuals.

After a few days of unsuccessful attempts, a contact puts us in touch with Jawad*, a young businessman from Kaniguram in the region of Lower Wana. He tells us how he experienced an attack from close by, during a visit to the town of Azam Warsak, a few hours' drive from his home:

> We saw a drone. It was too loud. It was not… I mean, we couldn't bear it. And after that, there was an attack from the drone by a rocket. […] Suddenly there was an extremely loud sound, like a screaming whistle, followed by a big bang. […] They targeted a single home, but

> because of that rocket almost 15 or 14 homes were destroyed. Their target was maybe five to six terrorists, but there were more than 20 people killed in that attack. And after 30 minutes, we, along with other people – there were so many people – we went there. And what we saw was pieces... of human bodies.

While this shocking experience forms the most prominent part of Jawad's account, it also concerns a singular event among many other experiences he has had of drone operations. In contrast, his recollection of the day-to-day UAV activity that he encountered on many other occasions is uncannily mundane. The drones came to Kaniguram in the evening or the night two or three times a week. Most of the time, they were invisible and only their sound could be heard, "which was similar to a small Fokker airplane". At first, people didn't know what the sound was. Only after they had learned through the media that they were drones, they would flee inside whenever they heard the sound.

The next day, I meet Hamid*, a middle-aged journalist who moved to Karachi a few years ago from a village in the Bajaur District, bordering Afghanistan. In 2006, the first large-scale drone attack took place in this region, killing 18 civilians at a mud compound in the village of Damadola (Williams, 2010). Hamid never experienced a drone attack himself though. Instead, he recalls how he heard the continuous sound of drones circling above his village, almost every night between 2006 and 2012, after which the operations have become less frequent.

Hamid says that unlike the sounds of the jet planes of the Pakistani Air Force, he usually perceived the sound of the drones as not unpleasant, even 'ordinary':

> The sound of the drones was like a flute, sometimes [it] would disappear and at other times it could be heard, as

> though someone was sitting at a distance playing the flute and you would hear the sound of it intermittently. [It] was not like a normal jet noise or a heavy sound that sets fear into people. It was an ordinary and distant sound, which is why some would not be so affected by the sound, but those civilians who lived nearby the militants they would fear it.

In the light of commonplace mediatized representations of drone warfare, these anecdotal accounts might seem somewhat surprising. The high-tech appearances and sounds of the former seem to fit neatly within the strategy of 'shock and awe': the conspicuous display of advanced and spectacular weapon power to intimidate enemies. While the – predominantly sonic – experiences of actual strikes with hellfire missiles, as described by Jawad, do seem to have this potential, the day-to-day occurences of the weapons are more ambiguous. Rather than the high-tech sounds of jet engines in promotional footage and popular culture, the sounds of hovering drones seem to be perceived mostly in the realm of the familiar and the everyday: somebody playing a musical instrument in the distance, the humming of a small civilian airplane flying by.

These sonic characteristics might be considered as what sound scholar Steve Goodman calls 'déjà entendu' in his book *Sonic Warfare* (2010): the "partial recognition of something heard but corresponding to the inability to attribute cause or location to the source of that sound effect" (p. 147). The sounds of hovering drones suggest everyday familiarity, while they do not immediately evoke a connection with the violent nature of the weapons that produce them. Through these 'déjà entendu' experiences, the presence of the threat of drones becomes connected to the ordinary sonic experiences of civilian everyday life.

Goodman uses the term 'affective tonality' for the "dimensions of mood, ambience, or atmosphere" that result from sonic experiences (p. 195). Unlike what we might expect when we consider the official representations of military UAVs, the affective tonality of drone warfare does not primarily seem to revolve around the spectacular emotionality of 'shock and awe'. Instead, its most common mode of perception concerns a tonality of the unseeming, the familiar and the banal.

The use of drones to perform military attacks within territories that are officially at peace blurs the boundaries between civilian and military space, both in a legal sense and in terms of lived experience (Chamayou, 2015 [2013]). The affective tonality of drone warfare further heightens this latter aspect. Through it, the soundscape of the everyday loses its innocence and becomes connected to imminent threats of violence. The drone starts to function as an everyday technology within a sonic ecology of dread.

* Names changed at the request of the interviewees for security reasons.

10. Smart Bombs, Bulldozers and the Technology of Hidden Destruction

In Western media, reports on the Gulf War (1990-91) have been dominated by high-tech weapons and equipment. In addition to the prominent featuring of ground troops using GPS (see also 'Positioning the Middle of Nowhere') and night vision goggles, video footage from cameras in the nose cones of 'smart bombs' showed how satellite and laser-guided navigation enabled hitting targets with high precision, thus supposedly preventing any collateral damage. The Gulf War was the first major military conflict in which satellite signals played a prominent role. Hence, the war has frequently been called 'The First Space War' (Anson and Cummings, 1991). Nevertheless, the conflict also had decidedly low-tech aspects. Armoured bulldozers and mine ploughs mounted on Abrams tanks were used to break through the so-called 'Saddam Line', a complex of trenches and mine fields along the southern border of Kuwait with Saudi Arabia. In addition to clearing landmines, the dozers and ploughs were used to cover Iraqi trenches with defending troops under vast amounts of sand (Sloyan, 1991).

The latter strategy only became publicly known after the conflict had ended. While the ground offensive took place, there were no media reports about it, probably because both journalists' movements and the material they were distributing were strictly controlled by military command (Deseret News, 1991). When details of the practice emerged half a year later, controversy arose over the legality of burying enemy combatants with sand.

Fig. 8: A D-7 armoured bulldozer of the US Army's 8th Engineers rides on a flatbed trailer as part of a convoy heading north during Operation Desert Storm. Photo: Staff Sgt. J.R. Ruark, 1991

The 1925 Geneva Convention, which the United States has ratified, specifies that the use of "asphyxiating devices [or] analogous liquids, materials or devices" (Hampson, 1993, p. 93) are prohibited. This may well apply to the use of sand to cover trenches with living soldiers in it. Also, the 1949 Geneva Convention requires "that the parties to the conflict shall ensure that burial is preceded by a careful examination with a view to confirming death". This is not possible when combatants are buried alive. Furthermore, the convention prescribes that weapons that cause "unnecessary suffering or superfluous injury" (ibid., p. 92) are unlawful. In this, it must be determined whether the military necessity of the action outweighs the suffering that is inflicted on the enemy. Notably, media statements by commanders of the brigades involved in the attack suggest that the primary reason for using the burial tactic was to minimize US casualties. Legal scholar Françoise J. Hampson points out that "it is far from clear that the desire to avoid military casualties amongst one's own forces is sufficient to determine military necessity". (ibid., p. 93)

It might appear that this rather crude and legally questionable strategy is diametrically opposed to the high-tech bombs that were proudly featured during press conferences in video presentations moderated by US General Schwarzkopf (rstahl, 2013). However, when considered from the perspective of the idea of 'clean warfare' and its representation, smart bomb and bulldozer have an important common characteristic. The camera footage from the nose cones of smart bombs, which became among the most iconic news media imagery associated with the Gulf War, showed how the weapons found their way to their targets efficiently and effectively. However, the footage always stopped at the moment of impact. Obviously so, because at this moment the camera was destroyed as well, but as a result the destructive impact of the weapon remained largely absent from its representation. In addition, as media scholar Roger Stahl (2018) argues, the 'weaponized gaze', from the perspective of the bomb, stimulates viewers to identify with the weapon, rather than the environment and people that are affected by its destructive power.

This erasure of the representation of physical violence is also a relevant aspect of the bulldozing tactic. When war correspondent Leon Daniel, who was part of the press pool embedded with the US forces during the invasion, asked an officer responsible for public affairs where the bodies of the thousands of disappeared Iraqi soldiers on the Saddam Line were, the latter responded: "What bodies?" (Sloyan, 1991). Burying the soldiers in their trenches made their bodies near invisible from the journalists who were led through the battlefield after the hostilities had ended. “The stench of urine, faeces, blood and bits of flesh”, (Sloyan 2003) which journalists had encountered and reported in earlier wars, were now absent. Similar to the way in which the smart bomb’s nose cone camera disconnects the perception of aerial bombardment from views of destruction, the burying of enemy combatants with bulldozers cleanses the battlefield of the perception of bodily harm.

Fig. 9: Video still from United States Air Force smart bomb nose cone camera footage, 1991

Conflict scholar Richard Bessel (2015) argues that – at least in most parts of the Global North – acceptance among the general public of violence and its representation has declined since the Enlightenment, especially in the second half of the twentieth century. Whereas in previous eras the use of violence for ends that were perceived as good was often broadly accepted (e.g. the use of terror by both Reds and Whites during the Russian revolution) more recent times have seen a rise in sensitivity to violence and its representation. Especially after the Second World War and – in the case of the United States – the Vietnam War, public opinion seems to have become increasingly critical of the use of violence by state institutions to pursue their interests, as well as the loss of citizens' lives in relation to this. The US government and military's apparent desire to represent the Gulf War as a clean affair, as well as resorting – behind the screens – to legally questionable methods of warfare to minimize risk of harm to their own troops, can thus be seen as an effort to appease public opinion at home and avoid loss of public support for the war effort as had happened over the course of the Vietnam War.

However, there is another aspect to this shared characteristic of smart bombs and bulldozers, which relates more directly to the experience of warfare by the combatants themselves. Unlike traditional 'dumb bombs', the violent impact of smart bombs is concentrated on a precisely defined location, which is shown on camera footage so that it might seem as if the observer is 'really there'. Yet, despite this, the operator of the weapon never comes face-to-face with an enemy combatant. The soldier who releases the smart bomb is never physically present at the site of impact. The situation of the armoured bulldozer or tank with mine ploughs is similar. Although in this case the driver of the vehicle is physically present on the battlefield, they remain inside their armoured shell, shielded not only from the enemy's gunfire, but also from any opportunity to meet their gaze.

This increased distancing between combatants – a further step in a development that arguably started with the first methods of mechanized warfare in WW1 – has implications for the ethical dimensions of warfare. Philosopher Emmanuel Levinas (2013 [1961]) suggests that the "rapport de face à face", the face-to-face encounter, forms the very basis for human sociality. He argues that humans' sense of ethics has its source on a precognitive level, in an embodied sensibility that takes places when we encounter the face of the other. Their face speaks to us and implores or commands us to discover our responsibility towards them. For Levinas, rather than the commandment 'Thou shalt not kill,' the source of our ethical potential lies in the encounter with the face of the other, which tells us: 'Do not kill me,' in other words: "Help me live, respect me for who I am, do not treat me as a thing" (Visker, 2014)

The account by a deserting German WW1 soldier of the development of friendly relations with enemy troops on the battlefield offers an example of the way in which a face-to-face encounter can unbalance the paradigm of violence between wartime adversaries. "At night time [the] troops were always standing together. Germans and Frenchmen met, and the German soldiers had a liking for that duty. Neither side thought for a moment to shoot at the other one; everybody had just to be at his post. In time both sides had cast away suspicions; every night the 'hereditary enemies' shook hands with each other" (anonymous, 1917, p. 144) While examples like these are rare and shouldn't be used to romanticize a war that was on the whole a gruesome excess of violence, the point of interest here is that there was a possibility for encounters like these. For the distant smart bomb operators and bulldozer drivers hidden inside their armoured vehicles the face-to-face encounters that underpin this possibility had become even less likely on the battlefield.

As such, the smart bomb and the bulldozer can also be seen as parts of a technological complex that prevents combatants from being confronted with the humanity of their opponents. While these weapons do reduce the risk of bodily harm to troops, the smart bomb and the bulldozer also play a part in preventing a fearsome scenario for the powerful elites whose interests they are ultimately defending: the chance that foot soldiers from both sides recognize their shared role as cogs in a machinery of power in which they may ultimately have little stake. In other words, the possibility that face-to-face encounters sow the seeds of a joint revolt of the disempowered. This happened on a few occasions during WW1, for example when Bulgarian frontline soldiers revolted and turned against the monarchy, stating that "Our enemy is not across the trenches [...] the real enemy is in Sofia." (Tsoneva, 2019). The technological paradigm of the Gulf War – both high- and low-tech – makes any dreams of an imperialist battlefield collapsing into a solidary revolt of the militarized proletariat against their oppressors, regardless of nationality, seem more distant than ever.

Image sources

Figure 8: <http://bit.ly/Ploeger24>
Figure 9: <http://bit.ly/Ploeger25>

11. Smart Technologies and Soviet Guns:

The dialectics of postdigital warfare

In March 2017, I travel to the frontline between Ukrainian forces and Pro-Russian separatist factions in the Donbass region, East Ukraine. While large parts of the frontline are usually quiet, unpredictable skirmishes and even heavy shelling regularly occur. I go to Shyrokyne, a destroyed seaside town in the area around Mariupol on the sea of Azov. Here, I am hoping to document uneventful frontline positions of the Ukrainian army to make a work about the everyday practice of alertness and waiting in warfare, which usually remains hidden in news coverage and public awareness outside conflict zones. However, contrary to expectations, I end up in a firefight while accompanying a small frontline patrol with members of the Ukraine Volunteer Corps, an organization affiliated with the nationalist organization, Right Sector.

Together with three soldiers, known under their noms de guerre as Bear, Carpenter and Steinar, I walk from a safe house on the edge of the no man's land to a frontline outpost opposite a separatist position, while installing a communication cable through an unfinished trench. Bear is the commander of the unit. The insignia on his camouflage baseball cap shows a picture of a cartoon-like bear holding a rifle. Carpenter is a jolly middle-aged man with a grey beard, wearing an old red and blue jumper under his open uniform jacket. Steinar, at least 25 years younger, has a much more sophisticated outfit. His

American-style battle dress fits perfectly and is complemented by a fashionable-looking camouflage-coloured Middle Eastern shemagh.

I film our journey with my smartphone, adopting the idiom of the medium through the inclusion of several selfie shots. This footage will later be transferred to 16mm film for my work *patrol* (2017). I am not the only one documenting though. The soldiers I accompany, as well as many of the others I meet at the safe house and the base camp, carry various state-of-the-art mobile devices with them: smartphones, GoPro action cameras, tablet computers.

Fig. 10: Bear, Carpenter and Steinar with GoPro camera and smartphone on the frontline in Shyrokyne, Ukraine, 2017

About halfway through our walk to the outpost, just after we entered the field that constitutes the no man's land, we hear automatic gunfire. "Down, down!" Bear yells. We duck down into the shallow, unfinished trench. As soon as the firing stops for a moment, we run forward towards the building that will provide cover. Bear and Steinar return fire through holes in the walls, while Carpenter collects a bazooka from the other side of the trench. Steinar fires with one hand while holding his GoPro in the other. Shortly afterwards, the camera is put into my hands and the men indicate that I should film them from behind a little wall. I document how Carpenter runs out into the open field next to the house with the bazooka on his shoulder, while Steinar and Bear follow him to give cover with automatic fire from their AKs.

Later on, after the firefight has quietened down and while we are on our way back to the safe house, Steinar picks up an unexploded mortar shell and enthusiastically proposes that we let it explode to have some fun. When I refuse to join and ask him to please put it down, he takes off together with Carpenter while I accompany Bear to join the other soldiers during their break. Shortly after, they return and proudly show us a video of the explosion on one of their smartphones.

Thinking back to this experience on the frontline, it feels like I was in a split temporality. In some ways, I seemed to have gone back in time. The soldiers' Soviet-era AK-74S assault rifles and their partly improvised uniforms remind me of film footage of ground warfare in Vietnam[9] and Afghanistan[10] in the mid-20th century. But this is contradicted by the electronic gadgets that featured prominently at the same time. The soldiers alternately shoot with mid-20th century weapons and contemporary audiovisual technologies. Thus, this war seems to be a kind of temporal hybrid where fighters simultaneously navigate a contemporary and an historical technological paradigm.

---

[9]See for example this news media footage of American soldiers in the Vietnam War <http://bit.ly/Ploeger10>

[10]See for example this footage from the Archive of the Soviet intervention in Afghanistan during the 1980s <http://bit.ly/Ploeger11>

My perception of a seeming contradiction in this juxtaposition of technologies is not that surprising though, considering that I have grown up in a media environment that has long emphasized the high-tech aspects of warfare conducted by European and North American nations. Since the Gulf War in the early 1990s, media representations of war have increasingly focused on the role of advanced technological weapon and navigation systems as I discussed in 'Smart Bombs, Bulldozers and the Technology of Hidden Destruction'. During operation Desert Storm in 1991, the US and allied forces introduced the embedded journalism system, where reporters' access to the battlefield was restricted to units and places designated by the military. In response to experiences with media reports that had represented US military operations unfavourably during the Vietnam War and which had had a significant impact on the loss of public support for the war effort in the United States, this media access regime was aimed at channelling media representations of the Gulf War in a way that matched the vision of military command (Keefe 2009). For example, it was effectively not allowed to show any imagery of dead allied soldiers, regardless of whether they would be identifiable. At the same time, the fact that access to the battlefield was heavily restricted meant that a lot of media reports were based on the press conferences given by General Schwarzkopf and other members of military command and accompanied by approved video material supplied by the military. In these briefings, there was a strong emphasis on the role of so-called 'smart bombs' and GPS-based satellite navigation. Video footage of bombs' nose cameras, showing unmanned targets, precisely locked within the cross hairs and subsequently neatly 'neutralized,' characterized much of the media representation of the conflict.

This focus on advanced technological systems in representations of war coincided with a technological momentum that took place in civil society around the same time. From the early 1990s, digital technologies – particularly personal computers – became a commonplace component of the domestic sphere in Europe and North America. Whereas from the 1950s until the late 1980s computers and other digital devices had largely remained the domain of specialists, large institutions and – from the late 1970s – dedicated hobbyists, they were now becoming part of everyday life. As such, the increasing prominence of technological systems in representations of Global Northern war efforts from the 1990s, through the Iraq and Afghanistan wars of the 2000s (after the 'smart bomb' of the Gulf War came the Stealth Fighter, the unmanned Predator Drone and other weapons innovations), seems to have moved in parallel with the spread of advanced digital technology-use in civil life.

However, from the 2010s a different development seems to have taken place. At the point where advanced digital devices – most prominently the smartphone – have penetrated almost all areas of everyday life in consumer culture, the high-tech perception of war efforts by nations in the Global North has started to become less pronounced. Whereas reports on the Iraq War (2003-11) initially showed a similar focus on advanced technologies to those in the earlier Gulf War, this interest waned towards the 2010s. Notably, during the US-led intervention in Syria (2014-) hardly any media attention was paid to the use of 'smart bombs' and other high-tech methods, despite the fact that they played a bigger role than in the Gulf War, during which – despite what media reports suggested – 92% of air to ground explosives used were actually conventional 'dumb bombs'. (Stahl 2018).

Here, Russell Davies' (2009) concept of the postdigital may provide insight. As I mentioned in 'Hi-Tech Everything: A report from the heart of techno-consumerism', Davies describes the postdigital as a (hypothetical) condition where society has been saturated with digital technology. On one hand, this would mean that digital technologies have become so commonplace and integrated in the everyday that their presence starts to lose prominence in people's perceptions. This is akin to Mark Weiser's (1996) concept of 'silent computing', which would entail computer technology and its interfaces becoming so smooth that they move to the background of experience. On the other, once digital technologies have become omnipresent, the oftentimes hyperbolic promises that were associated with them during their emergence tend to be reconsidered and nuanced. Once you actually live with advanced digital devices, it is easier to realize that although their capabilities are impressive they are not a magical solution to all problems.

In this context, we might consider Bear, Carpenter and Steinar as postdigital soldiers. Their use of weapons and strategies that have existed in this form for at least half a century is not unusual. Soldiers like them operated like this throughout the 1990s and the 2000s, despite the increasing focus on high-tech weapons in representations of the wars they fought in. However, what has changed is that high-tech audiovisual communication devices have now become so commonplace that even the unpaid volunteer foot soldiers of the paramilitary organisation I accompanied have access to them. As a result, the mediatization of conflict is no longer monopolized by 'official' journalism and large media organizations, which have often been subjected to far-going restrictions and limitations in their reports due to the embedded journalism system.

Now, soldiers can largely independently generate their own representations of the everyday of armed conflict, using smartphones, helmet cameras and other easily available gadgets. The footage they generate is not just shared among friends and family, but is also disseminated through informal online platforms such as YouTube where it becomes available to a broad audience. A quick online search brings up countless hours of battlefield footage by fighters all around the world, with all sorts of affiliations, ranging from US military to Daesh militias. This informal footage challenges the relative monopoly of so-called mainstream media outlets in the representation of war. Watching these contemporary soldiers operate in trenches, open fields and damaged buildings offers a counter narrative to the – often idealized and cleansed – representations of high-tech warfare that have been prominent since the early 1990s (see also 'Smart Bombs, Bulldozers and the Technology of Hidden Destruction').

Sanitized representations of high-tech war have played a significant role in the rise in popularity and spread of digital consumer technology, both in terms of military innovations driving technological developments in the civilian domain, and heightening the popularity of digital technologies in consumer culture. Global Positioning System (GPS) is a good example of this. As I discussed in 'Positioning the Middle of Nowhere', this technology became popular in civilian life in the wake of its prominent place in media reports on the Gulf War (Kaplan, Loyer and Daniels, 2013).

However, one of the outcomes of this process of cultural digitization – the omnipresence of mobile audiovisual media devices – has started to challenge the very narrative of technologization that arguably lies at the basis of its prominence in everyday life. We might call this the dialectic of postdigital warfare: when the technologization of culture through digital devices has reached the point of saturation – as the omnipresence of mobile smart technologies in the Global North seems to suggests – the wide availability of these technologies in themselves starts to undermine the belief in the transformative potential of technological innovation, first of all through its destabilization of the myth of clean and precise high-tech warfare.

Soldiers like Bear, Steinar and Carpenter provide us with images and sounds that operate within this dialectic. As such, they might help us to come to terms with existence in a postdigital condition, a condition in which digital technologies have become an omnipresent and taken-for-granted aspect of everyday life, while the promises of salvation through automation have given way to a continued awareness of the vulnerability and eventual death of our physical bodies.

However, while their footage might indeed offer a counter narrative to idealized and cleansed representations of high-tech warfare, it also easily facilitates a sense of nostalgia for a cult of masculine violence where protagonists stage themselves as heroic, individual warriors recognisable from Hollywood fiction.

Another disconcerting aspect of their documentary practice becomes apparent when we recall Susan Sontag's words about photography's potential impact on the events it documents:

> The omnipresence of cameras persuasively suggests that time consists of interesting events, events worth photographing. This, in turn, makes it easy to feel that any event, once underway, and whatever its moral character, should be allowed to complete itself–so that something else can be brought into the world: the photograph (Sontag, 1998 [1977]: p. 178)

When we regard the soldiers' simultaneous use of firearms and (digital video and photo) cameras from this perspective, a troubling picture emerges. The presence of their own cameras brings the risk that the acts of violence they perform become disconnected from moral concerns and primarily driven by a desire to generate 'good' footage. To what extent were Steinar's decisions to pull the trigger of his gun motivated by the gaze of his GoPro? Would Carpenter have fired the bazooka if Steinar hadn't brought his camera around to document it? Did my holding of their camera while he fired the grenade make me complicit in an act of war?

12. Techno-Mythology on the Border:

The pandemic risk society

A person entirely covered in a white plastic suit blocks the way of a man wearing a face mask. The latter slightly bows his head, after which a white device is aimed at his forehead by the person in the protective outfit. A few seconds later he walks on and the next person presents themselves to the device operator. At a placid pace, this process repeats itself, while a seemingly endless stream of people passes by.

This could well be a scene in a dystopian sci-fi movie where enslaved workers are registered upon entry into a factory. Or might it be a futuristic religious ritual, where the believer bows their head to an electronic relic that funnels the Holy Spirit? At the same time, the seemingly disinterested, yet focused way in which the operator aims the device at people's foreheads is somewhat reminiscent of the use of cattle guns in slaughterhouses (plapurdue 2008), akin to the notorious scene in the Coen brothers' film *No Country for Old Men* (2007).

Of course, the reality of the procedure described here is much more mundane, albeit nonetheless troubling. The device is an infrared thermometer operated by a border guard on the Polish-German border between Słubice and Frankfurt Oder, where I say goodbye to a friend on a Monday evening in the spring of 2020. The suit worn by the border guard is complemented by a face mask, plastic gloves and transparent eye cover so their body seems all but sealed off from the outside world. The thermometer enables the guard to accurately determine a person's body temperature by aiming an infra-red beam at their forehead, without touching them. When the registered temperature is above 37.7° Celsius, the person is considered likely to be infected with the SARS-CoV-2 virus, one of the symptoms of which is fever.

In this case, the person will immediately be taken to a mobile medical station that has been erected next to the border crossing, subjected to medical tests and subsequently put into quarantine. In a situation of feelings of insecurity and unknown threats, the operator appears as a beacon of reassurance and safety. While their protective suit gives the impression of a near absolute barrier from infection, their digital instrument produces empirical data on the basis of which a clear decision process is managed.

Fig. 11: Słubice border crossing, Poland, March 2020

In the early 1990s, sociologist Ulrich Beck suggested that, since the start of industrialization, threats to human existence like "famines, epidemics and natural catastrophes have been continually reduced" (Beck 1992, p. 97) to a point where these no longer form a significant hazard in people's everyday lives in post-industrial society. In their place, 'new risks' have emerged, which are the offspring of technological development, such as nuclear power, chemical and bio-technology. While the chances of occurrence may be smaller than with earlier, natural threats, the magnitude and longevity of destruction from hazards associated with these technologies is far greater; humankind has created technologies that are powerful enough to destroy its entire habitat.

For Beck, 'risks' are a distinctive subset of hazards that are based on "decisions that focus on techno-economic advantages" (p. 98). In pre-industrial eras, hazards appeared as a "stroke of fate" (ibid.) that came to society from outside. These hazards were perceived as apolitical and often associated with religious beliefs, e.g. a famine might have been considered a divine punishment. Since industrialization, however, the most prominent hazards have become those that are rooted in decision-making processes connected with responsibility and accountability. These risks are dealt with through a combination of observations and calculations on the basis of which precautions are taken and responsibility is attributed. For example, the hazards posed by a nuclear power plant are the offspring of human technology and are approached on the basis of calculations of probability, rather than merely accepted as "the dark side of progress" (ibid.) as would have been customary in a pre-industrial approach. Likewise, the risk of traffic accidents is determined on the basis of statistical data on past road accidents and vehicle and infrastructure design. In contemporary 'risk society', decisions on acceptable risks are then made based on a cost-benefit analysis, often expressed in monetary value. Beck argues that this principle becomes problematic in the case of risks where the hazard is total ecological destruction, as may be the case with a nuclear accident, because in such event there is no possible 'benefit' that could compensate for the 'cost'.

When we look at the hazards posed by the Covid-19 pandemic, a different realm of complications emerges though. The virus doesn't quite seem to fit within the division Beck draws. Its global spread, and thus its transformation into a serious threat, is facilitated by a technological infrastructure: aeroplanes and other means of transport that allow people to move easily far beyond their immediate surroundings. Thus, it seems to sit in the category of risk, decisions and precautions. However, the virus itself more closely resembles Beck's accounts of pre-industrial hazards posed by natural disaster, the kind of hazard that until recently appeared obsolete in the Global North. At the time of writing, little is known about how the virus spreads exactly, how contagious it is, what environments it thrives in, how many people have been infected and to what percentage of people it is dangerous. Also, in terms of its biological characteristics, the virus is – as far as we know – not the offspring of human technological development, and as such appears to be the kind of 'outside' threat, disconnected from the direct responsibility of specific people or organizations, which Beck associates with pre-industrial hazards.

This ambivalent characteristic is also reflected in the rather uncanny realm of political manoeuvring in connection with the Covid-19 pandemic. On one hand, the virus is framed within the paradigm of risk, at times with a determination that seems on the verge of desperation. Vast amounts of data, calculations and probabilities are published uninterruptedly, while expert interpretations and prognoses saturate media outlets. On the other hand, due to the vast number of unknown parameters, the pandemic often also appears as a natural disaster that is beyond politics or even reason. In day-to-day life, these two characteristics operate hand in hand. Government counter measures are introduced as careful decisions based on empirical data and calculation of probabilities, seemingly offering a considered balance between public health care and civil liberties. However, as soon as any doubt emerges about the basis of the calculated risk or the proportionality of measures introduced, the sentiment is shifted into the domain of the pre-industrial natural disaster, where regulations and restrictions are presented as commandments disconnected from politics that should be observed in a near-religious fashion (e.g. hikers on 'non-essential' solitary nature walks are sin shamed by UK police drones; Metro 2020). Thus, a worrying web of justification and accountability emerges where pretty much any decree that is associated with the pandemic can be introduced as simultaneously a carefully considered political decision *and* an inevitable and unquestionable threat response.

In this context, the wrapped-up border guard with infrared thermometer becomes a significant figure. The protective suit suggests an impenetrable precautionary measure where the chance of infection is all but eradicated. The thermometer collects empirical data, represented as a single number on its display, on the basis of which it is determined whether a subject poses an immediate threat: a near perfect instantiation of hazard management in the risk society. However, at the same time, the thermometer operator appears like a surreal being, an embodiment of hazard as mythological creature. It is this latter aspect that gives us some hints at what the realm of power and threat operating just beneath the narrative of calculation and control might encompass.

In the dystopian sci-fi movie scenario which the setup recalls, the hazard is oftentimes posed by autonomous technologies that have transformed from utilitarian support systems into a threat to humankind (The Terminator is a classic example). The technological aspect of the current pandemic threat follows a similar path: in a dialectical shift, global mobility seems to have transformed from a universal ideal of freedom to a technological threat that is beyond control.

At the same time, the ritual-like gesture of the procedure brings to the fore the thermometer's double role as a quasi-religious object. In the current state of affairs that is dominated by uncertainty and the unknown, measuring technologies – especially those that provide one-dimensional, digital information – become objects of hope, the scientific appearance of which suggests a lot more certainty than they actually offer.

But what might the thermometer's close aim at the foreheads of the passing testing subjects and its eerie reminiscence of cattle gun slaughter suggest? Maybe we should read this as a warning for another disquieting potential: in our desperation to find ways to control the virus amidst widespread uncertainty we can easily slip into the role of a complacent herd moving along the pathway of unquestioned state control.

13. Camera Surveillance and Barbed Wire

> The Concertina razor Wire design to provided with highly resistant blades, has a great capacity of penetration, while producing a deterrent to potential intruders. (sic)
>
> European Security Fencing[11]

CCTV cameras and barbed wire. Two technologies to control space, albeit in different ways. The camera allows for the monitoring of activities at a distance, and at the same time compels people to adjust their behaviour by giving them a sense of being observed[12]. Contrarily, barbed wire limits the movements of humans and animals in a direct, physical sense, while the imagination of physical pain evoked by the appearance of the material – an aspect frequently highlighted in product descriptions – gives it an element of deterrence. Whereas CCTV is a visual technology, barbed wire operates primarily in the realm of the haptic.

In Western-European everyday life, barbed wire is omnipresent and frequently used to fence off spaces from unwanted visitors. However, it is most commonly used in

---

[11]European Security Fencing (ESF) is the main supplier of concertina razor wire for border protection in Europe. <http://bit.ly/Ploeger29>

[12]See Michel Foucault's writing on the behavioural impact of the principle of the panopticon prison design in *Discipline and Punish* (1977 [1975]).

agriculture and landscaping to limit the movement of animals. Notably, for both applications, traditional twisted barbed wire is used almost exclusively. Historically, this type of wire has predominantly been used for agricultural purposes[13], unlike so-called NATO razor wire which has been developed specifically for use against humans. The dual purpose of twisted barbed wire means that it can still be read as a fairly benign technology of everyday civilian life, even when installed for the violent prevention of human movement.

Another aspect of interest in this context is the relatively inconspicuous presence of barbed wire in many security endeavours. It is not uncommon to find green-coloured barbed wire on a fence, to let it blend in with the environment and make its visual presence less prominent. The use of barbed wire in civilian spaces seems primarily focused on creating a practical physical barrier, rather than deterring through imagined injury.

For deterrence, camera surveillance has become the technology of preference. The prospect of being filmed while performing unwanted behaviour and having the footage used against you in legal action or social disciplining often works as an effective deterrent. However, this deterrence aspect loses its effectivity in most spaces of crisis, where the prospect of legal charges or social disciplining becomes futile: if one is considered 'illegal', the realm of the legal loses its relevance; in a condition of chaos, social discipline is impossible or insignificant. As such, the borders of the Schengen zone expose the practical ineffectiveness of the CCTV camera to enact a physical intervention. Even if you are detected immediately, this doesn't prevent you from moving quickly and disappearing

---

[13]Notwithstanding this, it was also applied on a large-scale in WW1 trench warfare and WW2 concentration camps.

before counter-action is possible, unless additional measures are taken.

This is where barbed wire reappears as a primary technology for spatial demarcation and separation. The promises of high-tech security – the holy grail of the post-Cold War Global North – collapse, and a seemingly archaic, low-tech approach is implemented in its full brutality: Endless stretches of 'concertina NATO razor wire' have been installed along the borders of Hungary, Slovenia, and Spain, among others. Razor wire threatens the observer with its display of perceptibly sharp blades. It won't just poke a little hole in you. It will entangle you and slash you open. Moreover, its association with the realm of the military, its specific design to hurt humans, removes the sense of positive ambiguity of the agricultural wire that is used in everyday security fencing.

Foregrounding the realm of the haptic, the domain of bodily pain, either in fearful imagination or in actual, physical experience, razor wire pushes the human subject toward its base faculties. The reflective experience of self-awareness and observation evoked by the CCTV camera is replaced by the immediacy of bodily fear and pain. Here, the words of Elaine Scarry are useful: "Physical pain does not simply resist language but actively destroys it, bringing about an immediate reversion to a state anterior to language, to the sounds and cries a human makes before language is learned." (Scary, 1985, p. 4)

Civilization ends at the razor wire fence.

14. The Smart Fence is the Message:

EU border barriers as violent media

Since 2015, hundreds of kilometres of reinforced barriers have been built on the outer borders of the European Union. These fences, erected to prevent migrants from entering the EU, largely consist of mesh fencing and so-called NATO razor wire, but are often also equipped with loudspeakers, lights, electric shock devices, and various sensors. The Hungarian anti-immigration fence on the border with Serbia, often dubbed a 'smart fence,' features a particularly broad range of such technological additions.

On the first days of winter in the end of 2018, I stand directly next to this fence, close to the city of Subotica in Serbia. However, my experience is not one of technological awe. The barrier does feel intimidating and threatening, but this is first of all because of its size in relation to my body and the fact that my eyes are drawn to the sharp points of the razor wire blades. The barrier consists of two parallel fences over four metres high, separated by a patrol road, with a total of eight rolls of razor wire installed on it, stretching out as far as I can see in both directions. Despite its high-tech promises, the fence gives the impression of a crude physical obstacle, violently separating spaces and people.

Fig 12: Smart fence. Dani Ploeger, *European Studies no. 1* (2019). C-Type print on paper, 6x6cm

As discussed in 'Camera Surveillance and Barbed Wire', razor wire is primarily conceived to communicate a threat of physical harm, rather than to inflict it. According to the manufacturer of the razor wire used in the Hungarian fence – European Security Fencing in Malaga, Spain – it is a 'passive safety' product that primarily operates as a 'deterrence.'[14]

There are other deterrent aspects to the fence as well. Yellow warning signs tell you that it is electrified and touching it poses a shock hazard. There seems to be no actual danger though: the shock is reportedly only 'mild.'[15] If you do end up touching the fence after all, a sensor system activates the loudspeakers, which play a voice recording of a warning message in various languages, pointing out that damaging or crossing the fence is a criminal act under Hungarian law. Meanwhile, border guards observe you on camera and intervene if they deem this necessary.

---

[14] <http://bit.ly/Ploeger29>

[15] <http://bit.ly/Ploeger30>

This is what happened when I stole a piece of razor wire from the fence. Shortly after I started cutting the wire, two border guards arrived by car and started yelling at me. I ran off as fast as I could, scared to be pepper sprayed or otherwise assaulted (my video *Border Operation*, 2018, documents this action).

Considering all this, it becomes clear that instead of looking at the Hungarian border fence as a more-or-less technologically enhanced obstacle, the following might be a more accurate reflection of the way it operates: an "intervening substance through which sensory impressions are conveyed or physical forces are transmitted". This is one definition of a 'medium.' (Oxford Dictionary of English, 2019)

EU-border fences are media that transmit different content, ranging from notions of danger, (il)legality and authority, to affective exchanges between people protecting and challenging its structures. However, if we follow McLuhan's (1964) adage that 'the medium is the message,' another question arises. If the medium itself – rather than the contents it transmits – defines its broader cultural relevance, what would the 'message' of these EU border fences be? The first answer that comes to mind seems straightforward: the broader relevance of these fences is their limiting effect on migration. But this answer is less conclusive than it might appear. It remains debatable whether the new border fences have actually led to the significant decrease in the influx of migrants to Europe that has taken place in the years since they were built.[16] Moreover, when we examine

[16]Various critics have argued that other factors, such as the evolution of the conflicts in Syria and elsewhere and a limit to the total number of people who are able to flee, played a more prominent role in the decline than the border fences. Besides, it has been pointed out that erecting barriers tends to mainly lead to people diverting to

the way the fences – particularly the Hungarian one – have been represented in news reports and government communication another message comes to the fore.

A 2017 press release on the website of the Hungarian government reads "'Smart fence' is working, second border barrier is being erected".[17] In addition to announcing the supposed success of an experiment with a new fence design in preventing people from crossing the border, the title already shows that another interest is at stake in the article as well. The use of the adjective 'smart' is significant. The designation 'smart' is conventionally used for everyday technologies that enhance comfort and efficiency through detailed and accurate data collection and processing. In other words, by referring to the border fence as 'smart' it is framed as an innovation connected to the world of high-tech consumer culture. This focus on the technologically advanced aspects of the fence has also featured prominently in media reports, often accompanied by images that put the fence's sensor elements at the centre of attention.[18] [19]

Thus – especially when considered from the perspective of its representations – we could say that the '*smart* fence' is the message. While digital technologies make up only a small part of the fence they tend to be foregrounded in representations through the use of the idiom of digital consumer culture, combined with a visual focus on – supposedly – advanced technological features. Thus, the fence contributes to the idea that EU border protection is part of the everyday paradigm of

---

alternative routes, instead of refraining from migration. See for example: <http://bit.ly/Ploeger31>

[17]<http://bit.ly/Ploeger33>

[18]<http://bit.ly/Ploeger32>

[19]<http://bit.ly/Ploeger34>

digital consumer culture, rather than the realm of territorial, military violence. The fence does indeed have a few features that involve digital sensor technology, but the framing of the whole structure as a 'smart fence' or 'high-tech border' draws attention away from the fact that its main components are razor wire, mesh fencing, metal bars and a 'vehicular access trench'[20] in between two parallel barriers. Rather than a high-tech innovation, the fence actually isn't very different in design and function from the Iron Curtain that separated Europe during the Cold War. In the end, its main modes of operation revolve around the impression of physical threat evoked by the razor blades and metal obstacles it is made of, and the physical obstruction it poses. It has little to do with the conventional associations of 'smart' technology with an optimized, 'civilized' everyday life.

This framing of technologies of violence as clean precision instruments is by no means an anomaly. It is symptomatic of a broader cultural practice that uses narratives of technologization to justify means of violence, such as 'smart bombs' and armed drones. In other words, there is more at stake than just this fence. This is about a fundamental culture of violence at the basis of techno-consumerism, which operates through a logic of forging acceptance for technological militarization through the idiom of everyday high-tech gadget culture.

---

[20] <http://bit.ly/Ploeger33>

15. The Deluxe Anti-Terrorist Barrier

In front of Notre Dame cathedral in Paris, I sit down on one of the solid granite blocks that are placed in various spots around the square and on the adjacent pedestrian roads. Its surface is smooth, edges rounded. Almost elegantly, its material texture and colour match the typical Parisian kerbs. Tourists and other passers-by sit on these objects when the classic wood-and-iron 'Davioud' benches on the bridge are occupied. Others use them as a step to take pictures, or a plinth to pose on. Despite their appropriation as urban furniture, the positioning of the blocks also suggests another function though. During the day, a police car blocks the only opening in the line of blocks on one side of the square, while during special events a bright yellow so-called 'mobile anti-terrorism barrier' is installed across a gap in the arrangement on the opposite side. It is hard to overlook these obstacles' function as a security measure to prevent trucks and cars from driving into the crowds of tourists lingering in front of the cathedral.

In this respect, their purpose is similar to many of the rough concrete barriers with lifting holes and metal hooks that can be found around buildings and squares elsewhere in the city, and they fit into a broader set of government-led 'Vigipirate' security measures including metal detector gates at public buildings and the frequent presence of armed military personnel on the streets. Most of these measures, which were installed in the direct aftermath of the terrorist attacks in 2015, seem likely to constitute a clear rupture in the everyday experience of the urban landscape, evoking an immediate sense of threat and emergency, or even a state of war.

The granite blocks around the Notre Dame are more ambiguous though. They can be simultaneously perceived as public furniture *and* anti-terrorism measure. As such, they seem characteristic of changes in the organization of public space that have persisted beyond the end of the state of emergency, which lasted from 2015 until 2017. This deluxe successor to the concrete, temporary anti-truck roadblock blends into its urban environment and is widely embraced by users of public space as an object of convenience. However, its conspicuously strategic positioning, as well as its shape, which is reminiscent of its concrete predecessors, mean that it never becomes entirely dissociated from its place in the paradigm of security and threat. These objects act as half-conscious reminders of a condition of fear and threat that is lurking just below the surface of the cheerful hustling and bustling of the touristy everyday of this site.

Fig. 13: Anti-terrorist barriers at Pont au Double, Paris, 2018

As such, they fit in with a broader cultural development in our engagement with security and threat. In *The Administration of Fear*, Paul Virilio (2012) discusses how the proliferation of digital news media over the past two decades has led to the distribution of more news events at ever increasing speeds. Through 24/7 news feeds, updates on the latest incidents all over the world are disseminated almost instantly and continuously. Instead of facilitating a more detailed assessment of events and thus triggering empathetic action on the side of the recipient, the fast sequence of detailed mediatizations of catastrophes and threats paradoxically promotes a permanent state of fear. Whereas, in previous media eras, fear used to be 'related to localized, identifiable events that were limited to a certain timeframe', it has now increasingly become 'an environment, a surrounding, a world'(p. 14) .

In this context, the ambiguity of the granite block is significant. Through its function as a traffic barrier it suggests the possibility of a terror attack, but unlike its temporary, concrete and metal predecessors it is simultaneously experienced as an ordinary, everyday piece of street furniture. With this deluxe anti-terrorism barrier, a sense of impending catastrophe has become fully integrated in the grid of the urban everyday. As such, it can be seen as a literal, material manifestation of the everyday environment of Virilio's administration of fear.

16. Struggle and Expand:

The Delta Works as colonial technology

> The story of the Delta Works is grand and marvellous. Fourteen dams protect the vulnerable Dutch delta against storm surges. [...] The delta forms the cradle of our culture, the habitat for millions of people [...] 'The Dutch can do it' is what the Delta Works express [...] Thus, the Works have a much bigger meaning than merely protecting against a dangerous sea. They form a crucial link in the international reputation of the Netherlands as a country of water engineers. (Steenhuis, 2016)

A few hundred metres from my home in the Dutch harbour town of Flushing, enormous dikes prevent the sea from flooding the streets at spring tide. They were reinforced and enlarged to their present size as part of the Delta Works, a vast complex of dams and dike improvements in the province of Zeeland and adjacent areas in the south west of the Netherlands, constructed to protect the land in the Scheldt delta from flooding. The government-led project, largely funded with income from the national gas reserves, was initiated after the flood of 1953, which affected a total area of 200,000 hectares in the region and led to the death of over 1,800 people and almost 200,000 cattle and poultry.

Usually, the Delta Works are presented as a project that has been necessitated by the unpredictability of nature, a human-made protection against an inevitable threat. However, looking at the historical development of the region, it becomes clear that this is only a partial truth, at most. Before dikes were built, the Scheldt delta consisted largely of wetlands that would flood regularly and which continuously changed shape. From the middle-ages some of the tidal areas were enclosed by dikes to enable permanent inhabitation. Draining of water from peat grounds due to agriculture and salt and peat extraction in these areas then resulted in the sinking of land that was previously above sea level. In addition, up until the 20th century, more and more land was reclaimed to further expand agricultural production. This, in turn, necessitated the erection and reinforcement of ever more dikes.

Thus, the Delta Works should also be seen as a response to problems that resulted from preceding agricultural endeavours. Especially from the 17th century, when traders of the Dutch East India Company started investing in large scale land reclaiming projects (Sikkema, 2014), the water works in Zeeland started to operate as what Raj Patel and Jason W. Moore call a 'frontier' in their analysis of the history of colonial capitalism: "a site where crises encourage new strategies for profit [...] the encounter [zone] between humans and all kinds of nature [aimed at] reducing the cost of doing business" (Patel and Moore, 2017). In this context, the Delta Works may be considered a frontier technology to overcome yet another challenge in a long tradition of securing access to cheap and profitable natural resources.

In addition to a frontier technology – a means to an end – the Delta Works have more recently also become an end in itself, as a commodity in a globalized service economy. The Dutch government, universities and businesses promote the water engineering technologies developed during construction and maintenance of the Works as an export product, using the dams as a real-life showroom for international delegations interested in purchasing Dutch engineering skills and knowledge (Dikkenberg, 2012).

The coat of arms of the province of Zeeland features a lion emerging from the waves, accompanied by the motto 'Luctor et Emergo', I struggle and emerge. The Delta Works are oftentimes celebrated as a high-tech manifestation of this supposedly typical regional attitude – 'The Dutch can do it' – and they play a significant role in Dutch cultural awareness and identity. However, to be truthful, we might need to add a second motto: 'Luctor et Expando,' I struggle and expand.

Postscript:

Artificial techno-myths

The sixteen essays that make up this collection engage with a broad range of devices, places and cultural practices, many of which may seem to have little connection to each other. What does the sound of a Predator drone in Pakistan have to do with a second-hand computer monitor on a market in Nigeria? How is an infra-red thermometer during the Covid-19 pandemic comparable to a mobile phone in the Donbass War? What does an anti-terrorist barrier in front of Notre-Dame in Paris have in common with a strap-on dildo in a sex shop in London's Soho? They may not share much more than the underlying theme I outlined in the Introduction: all are in some way concerned with practices and representations of technology outside the realm of use and perception that was envisaged by their developers and marketers.

However, through this rather broad shared characteristic, all of these case studies – if we can call them that – are connected to another thematic element as well: they examine technological devices and practices that evoke tensions, conflicts and collisions with what philosopher John Gray has described as the 'myth of progress', a widespread cultural belief that "new technologies will conjure away the immemorial evils of human life" (Gray, 1999). Through this myth, technology is primarily associated with notions of innovation, immateriality and uninhibited growth, while ignoring its embeddedness in processes of social domination and intrinsic connections to the material realm of resource harvesting and environmental conflict.

In the terminology that surrounds the so-called 'smart fence' on the EU border and the popular cultural sonic representations of Predator drones, technologies of violence are rendered harmless through the invocation of imaginaries of the high-tech. Once a technology of violence is framed as a state-of-the-art innovation its troubling brutal dimensions seem to move to the background. In a similar vein, the conspicuous use of infra-red thermometers during the Covid-19 pandemic appears to offer reassurance in a situation dominated by unknowns through the association of digital precision instruments with an assumed intrinsic power to know and control the world around us. Meanwhile, the relatively low-tech characteristics of the X-Tigi 'dumb' phone and its explicit connections to the physicality of its surroundings emphasize the extent to which the smooth and light characteristics of its top-end, smartphone counterparts convey a promise to transcend the difficulties of material existence through technologization. The exclusion of dust from representations of everyday computing is significant in this way as well: this unruly substance does not fit in the vision of digital technology as a means to magic up an organized world.

It is in this mythical dimension that I see the relevance of studying technologizations of the everyday through a multitude of fragmented reflections, rather than a focused, linear approach. In his study of myths in French culture during the 1950s, Roland Barthes suggests that a myth's 'concept', namely the ideologies or beliefs it conveys, is "not at all an abstract, purified essence; it is a formless, unstable nebulous condensation" (Barthes, p. 118). Accordingly, the myth of progress does not have an essence or core that can be unravelled or grasped through an in-depth analysis of a particular technological artefact or practice. Rather, it is like a hydra, the meanings and consequences of which shift in relation to the circumstances under which it operates. Especially on the 'frontiers of capitalism' (Patel and Moore, 2017) – those contexts in which the processes of capitalist production and consumption are expanding or receding – it appears to manifest itself in ever surprising ways and forms.

However, this does not mean that it is not possible to home in on the nebula of meanings that constitute this monster. By building a multi-faceted body of knowledge that is deliberately inconclusive, open-ended in the number of its parts and which does not shy away from internal contradictions, we can gain insights into the tendencies and potentials of this myth, which now seems to permeate contemporary everyday lives to various degrees almost anywhere in the world. This collection of texts, which are deliberately seeking out the peripheries of techno-consumerism in a wide range of cultural contexts, will hopefully form a modest contribution to this field of knowledge.

This does not address the elephant in the room, though: how can a variegated understanding of the myth of progress that would emerge from this diverse body of knowledge pose an actual challenge to the exploitative and alienating dimensions of contemporary techno-consumerism? Identifying problematic elements through cultural analyses, as the texts in this collection have sought to do, will in itself hardly have actual transformative potential. Barthes addresses this difficulty by pointing out the limitations of the theoretical analysis of myths. He suggests that the analysis of a myth is done through 'metalanguage', a form of speech that will always remain separate from the mythical realm itself. As soon as one starts to 'decipher' a myth, one is cut off from those communities that are invested in its promises. As a consequence, a theoretical analysis in itself is unable to challenge the myth in the space within which it operates, it will always remain on the sideline. In order to pose a concrete challenge to myth, Barthes proposes a different approach: "The best weapon against myth is to mythify it, and to produce an *artificial myth*" (p. 134).

This is how the essays in this collection connect to my work as an artist. The artwork I have made over the past decade has evolved in conjunction with the critical reflections in these texts. Through a continuous interplay between personal experiences, practical explorations and theoretical inquiry, I approach the artistic process as an attempt to create artificial myths in the sense of Barthes' call: alternative technological imaginaries that seek to defy the excesses of the myth of progress. I implanted electronic waste in my body to turn myself into a waste cyborg that opposes the popular cultural image of the high-tech cyborg as a harbinger of progress (Electrode, 2011), I made a sci-fi short film in Kenya that features repurposed electronic waste as constituents of a utopian high-tech world (The Cults, 2020-21) and I appropriated an industrial stress testing machine to exhibit the destruction of a brand new high-tech shaving device (Laboratory of Electronic Ageing, 2019), among others. A more comprehensive overview of artworks that emerged in concomitance with the essays of this collection can be found in the acknowledgements.

It is through this kind of intertwinement of theoretical reflection and creative practice that I hope a continuous interplay can be established between critiques of the myth of progress – in a wide variety of its 'unstable' and 'formless' manifestations – and emerging clouds of challenging, artificial techno-myths. The ideological noise that this cacophony of shifting myths and counter-myths would generate might then form the foundation for another, more radical and absolute response to the status quo of globalized techno-consumerism.

Barthes also identifies a second method to defeat mythology: revolution. As an act to overturn the socio-cultural status quo, revolution announces itself openly as exactly that what it is. As such, it excludes myth altogether, and enables its destruction. If we really want to enable life based on sustainable interrelationships between people and the world we live in, radical intervention may be the only viable possibility left. Considering the impending global environmental catastrophe, which is closely tied to the obsession with eternal growth and expansion that is accommodated by the myth of progress, the time of resolving issues through a tweaking of existing dominant models has passed. What the techno-cataclysm we should strive for might look like – in theory and in practice – and how its material conditions may be brought forth could be the subject matter for a future project.

Acknowledgements

The texts in this book would not have existed without the support of numerous people in all the places I have visited over the past years. I would like to thank Greenman Muleh Mbillo and Joan Otieno, with whom I chased old and new electronic gadgets across Kenya; Jelili Atiku for taking me along electronic-waste dumps and recycling sites in Lagos, Nigeria; Zahra Stardust for inviting me on a kinky shopping trip to London's Soho; Janet Chan, who enabled me to participate in e-waste recycling in a Hong Kong factory; Mehreen Hashmi and Yasir Husain for supporting me with the interviews in Karachi; Dima Kolchinsky, my fixer in Ukraine, who cunningly navigated us along the frontline; Julia Harasimowicz, the curator of my residency in Warsaw during the pandemic; Gabriela Seith for accompanying me to the border fence at Subotica in Serbia; Bénédicte Alliot for the conversations about urban terrorism during my residency at the Cité internationale des Arts in Paris; Aseel al-Yaqoub for her proxy navigation of the Kuwaiti desert; and Hava Carvajal and Max Packham-Walder for throwing a unicorn off the Eastern Scheldt Storm Surge Barrier with me.

I am also indebted to The Royal Central School of Speech and Drama, University of London, Leiden University, V2_Lab for the unstable media in Rotterdam and ruimteCAESUUR in Middelburg for their support of my work. In addition to my own funds, the research presented in this book was financed through research grants from the Arts and Humanities Research Council, Global Challenges Research Fund, Newton Fund and Mondriaan Fund. It was a joy to work with Andrew Carey from Triarchy Press, whose comments on the manuscript have been greatly inspiring.

1. Tactical Transgressions: Bashar al-Assad's phone

This text was originally written for a lecture I gave at Aarhus University, De Montfort University Leicester and the University of Greenwich in 2019 and 2020. In the context of its engagement with mobile phone bomb triggers, I made *Trigger*, 2020, an installation with a Nokia 105 phone, gold plated C106D-Thyristor and smoke machine. The work reprocesses documents and artefacts related to terrorist Ramzi Yousef's iconic bomb trigger design from the 1990s, which involved a C106D thyristor connected to a digital Casio watch. Its principle is still frequently used to create mobile phone bombs. In the installation, Yousef's design is extrapolated as a perverse effect of the society of the spectacle. A phone equipped with a gold-plated thyristor is attached to a smoke machine. When the number is dialled, the event is set off.

2. E-Waste in Cling Film: The Symbolic Order of Technological Progress

An earlier version of part of this essay was published in *Leonardo* 50(2) in 2017. It was written in conjunction with my performance installation *Recycled Coil*, 2014, commissioned by transmediale festival in Berlin. For this work, parts of a deflection coil from a discarded CRT television collected in Lagos were installed on my abdomen. A body piercer sewed magnetic wire from the coil through my skin and attached the coil's connector above my belly button. For the duration of the exhibition at transmediale, an electric current was run intermittently through the coil. Thus, an electromagnetic pulse signal was generated. For several hours a day, I presented myself in the exhibition space, accompanied by a magnetometer detecting my magnetic field.

3. Hi-Tech Everything: A report from the heart of techno-consumerism

This report first appeared in *Body, Space and Technology* 17(1) in 2018. It formed the basis for my installation *Laboratory of Electronic Ageing*, 2019, which I made as part of my research fellowship at Leiden University in the Netherlands, 2018-20. The work combines an industrial drop testing machine, electric shaver, video monitors, loudspeakers, vinyl laboratory floor, fluorescent lights and found objects.

The Philips Series 9000 Shaver is marketed as a high-tech gadget that is subject to rapid innovation and obsolescence. Nevertheless, it originates from and ends as bare materiality. The industrial 'drop testing machine' – normally used in manufacturers' laboratories to test product durability – makes this visible in the work. In an endlessly repeating simulation of a user accidentally dropping the device, the high-tech shaver is gradually transformed into a state where it is no longer has any functionality for humans, while it remains present as material.

4. Eerie Prostheses and Kinky Strap-Ons: Mori's uncanny valley and ableist ideology

The initial version of this text was published in *Body, Space and Technology* 13(1) in 2014. The unrealized work mentioned in the introduction is *Strap-On*, a telematic performance with fucking machine, live video and motion tracker in collaboration with pole dancer, porn actress, and feminist activist Zahra Stardust. In the performance, I would watch an erotic webcam performance by Zahra. However, rather than merely watching a commonplace webcam performance, the male observer's body in this performance is simultaneously the target of the actions of a robotic device that is controlled by the female performer's movements, which are tracked by a sensor and subsequently transmitted over the internet. Thus, the work challenges the widespread simplistic perception of heterosexual erotic performance as a practice where a male spectator's gaze 'objectifies' the female performer.

5. The Dirt Inside: Computers and the performance of dust
I wrote this text for the annual Digital Research in the Humanities and Arts conference in 2015. Connected to the text, I exhibited a ready-made, *ELECTRONIC 506 (garbage cyborg)*, 2015. The work is a cyborg composed of an obsolete vacuum cleaner and household dust.

6. Frugal Phone / Material Medium
This reflection is based on an unpublished note, which I wrote on the way back from Nairobi to Europe sometime in 2019.

7. Orodha: The Ultimate Fetish Commodity and its Reversal
This text was originally written at the invitation of Andreas Broeckmann for his event *Les Immatériaux* at V2_Lab for the unstable media in 2019. The slide projector discussed in the text formed the starting point for my short film *The Cults*, 2020-21. This 16mm film takes an afro-futurist perspective on the appropriation of obsolete consumer technologies and their transformation into devices with new uses and meanings, a commonplace practice in Kenya. Drawing from the style of

mid-20th century ethnographic films, it imagines an alternative technological culture that takes it starting point from East-African stories, myths and practices in order to challenge the standardized technological narratives of globalized consumer culture. The film reworks a 1930s ethnographic text by a British colonial administrator on the early 20th century Mumbo cult, an anti-colonial religious movement. Taking fragments of this colonial text out of context and thus reversing its meaning, it is used to critique neo-colonial resource-harvesting and the fetishization of standardized consumer technologies.

8. Positioning the middle of nowhere: GPS technology and the desert

Commissioned by the Kuwait National Pavilion at the Venice Architecture Biennale, this essay first appeared in the catalogue of the 2021 pavilion, *Space Wars*. It formed the context for my work *A Space War Monument*, 2021, which was exhibited as part of pavilion (see 10. below). However, the initial idea for the text originated from my project *A New Middle of Nowhere*, which I realized during a residency at the Ujazdowski Castle Center for Contemporary Art in Warsaw.

A New Middle of Nowhere, 2020, uses emulated satellite signals to enable a new experience of being in 'the middle of nowhere' in the era of omnipresent GPS location tracking. I commissioned a mathematician to calculate the precise geographical centre of one of the last remaining primeval forests in Europe: Białowieża National Park in Poland. Subsequently, I created an object that emits a GPS spoofing signal with these coordinates: the radio signals from actual satellites are overpowered by emulated signals that are broadcast at the same frequency. This results in GPS devices in its vicinity indicating a geographical location that doesn't correspond with the device's actual location. I then undertook a clandestine foot journey into the Strict Reserve of Białowieża Forest to reach the geographical centre of the National Park. There, the GPS spoofing object was temporarily installed. While it was transmitting, any GPS-enabled devices in the area would have indicated the exact centre of the park, regardless of their true location: a middle of nowhere 2.0.

9. Sounds of Violence: The affective tonality of high-tech warfare

A preliminary version of this text appeared in *Body, Space and Technology* 18(1) in 2019. It formed the basis for my work *Sounds from the Air*, 2018-19, which consists of a 360-degree 3D video and three interactive objects. The video combines vocal impressions of the sounds of drone operations by the interviewees mentioned in the text with digital video animations and satellite imagery. The three interactive objects are heart-shaped electronic sound boxes intended for speaking teddy bears, re-painted in grey, drone camouflage colour; when you press the hearts, the voice recordings of drone sounds are played.

10. Smart Bombs, Bulldozers and the Technology of Hidden Destruction

Originally written for the catalogue of the Kuwait National Pavilion at the 2021 Venice Architecture Biennale, this essay has thus far remained unpublished. It was deemed unsuitable for the catalogue by the curators, among others because of its explicit references to potential war crimes committed by the United States, and allied nation of the State of Kuwait. The artwork I created for the pavilion, *A Space War Monument*, fuses the two technologies that are central to the text: bulldozers and GPS navigation.

A Space War Monument consists of video documentation of a land art performance in which a 100 by 100 metre square area in the Arabian Desert is flattened. Using a GPS-controlled bulldozer, remains of 'Saddam Road' – built my Iraqi troops during the war – are removed within this area. A machine that fuses two prominent Gulf War technologies is used to erase traces of the conflict. An act of erasure to remember the unknown bodies that disappeared under the sand.

11. Smart Technologies and Soviet Guns: The dialectics of postdigital warfare

Part of this text previously appeared in the *International Journal of Performance Arts and Digital Media* 15(2) in 2019. The 16mm film mentioned in it is *patrol*, 2017. I transferred my smartphone footage of the firefight discussed in the text to celluloid film in order to connect the mediality of the work to the era of the weapon technologies represented. This engagement with the intersections of digital culture and low-tech warfare technologies was also central to *artefact*, 2017. It involves a wooden handguard from an intensively used Kalashnikov assault rifle, which is displayed like an archaeological artefact. Abrasion of the varnish shows the points of contact with the hand of the previous owner. A monitor shows a high-resolution 3D scan of this object inserted into a slick digital model of an AK-47. The weapon rotates against a standard backdrop of a blue sky reminiscent of weapon selection menu in a game.

12. Techno-Mythology on the Border: The pandemic risk society

I wrote this as a note accompanying a short video I shot at the Słubice border crossing in Poland. In response to the sci-fi reminiscence the situation evoked in me, I put together a little video homage, *Four Months After Blade Runner*, to Ridley Scott's 1982 film (after a book by Philip K. Dick), which plays in November 2019. Both my text and the accompanying video were originally merely intended as a leisurely sketch and remained unpublished until now.

13. Camera Surveillance and Barbed Wire

This text was written for an exhibition at ruimteCAESUUR in Middelburg, Netherlands, in 2018. Among others, the exhibition involved installing a razor wire barrier across the gallery entrance together with artist Peter Puype.

14. The Smart Fence is the Message: EU border barriers as violent media

This essay is based on a lecture I gave at Kunstraum Kreuzberg/Bethanien in Berlin in the context of the exhibition 'Up in Arms', which included my installation *OUR VALUES*, 2019, commissioned by the Neue Gesellschaft für bildende Kunst Berlin.

OUR VALUES engages with razor wire produced by European Security Fencing (ESF), a factory based in Malaga in Spain. ESF's wire is used in many of the newly reinforced outer borders of the European Union, including the Hungarian-Serbian border. The euphemistic and positive-sounding corporate language used on the ESF website – which seems to have been translated with Google Translate – and the possibility to order razor wire 'quickly and safely' in the online shop suggest that we are dealing with an everyday consumer good, provided by an ordinary business in a global market economy. This forms a stark contrast with the archaic, violent reality of the product in its application as part of Fortress Europe. In *OUR*

VALUES, ESF wire is attached vertically to a metal stand with a horn loudspeaker, similar to the loudspeakers installed on the Hungarian border fence. Through the loudspeaker, the Google Translate voice monotonously speaks the corporate values that are listed on the ESF website. In August 2019, I travelled to Serbia to install this object next to the Hungarian border fence.

15. The Deluxe Anti-Terrorist Barrier

In the autumn of 2018, I stayed in Paris for a residency at the Cité internationale des arts. I made an installation, *Aftermath*, based on an examination of the remains and traces of the state of emergency in Paris, which lasted from the Charlie Hebdo attacks in 2015 until the end of 2017. What stays after the initial state of high alert has ebbed away? What characterizes the post-emergency mundane? I collected promotional videos of mobile anti-terrorist roadblock manufacturers and media footage of terrorists' private spaces that are floating around the internet. I turned my studio at the Cité internationale des arts into a gamers cave – an uncanny parallel universe to the secluded private space of the lone wolf terrorist – and made a truck racing game with 3D scans of roadblocks next to the Notre Dame cathedral. The text formed part of the catalogue accompanying this work.

16. Struggle and Expand: The Delta Works as colonial technology

This text was originally included in the video installation *COZY LAND*, which I made together with performance artists Hava Carvajal and Max Packman-Walder in 2018. During a short residency at ruimteCAESUUR in Middelburg, NL, we wandered along the Delta Works and the remains of the

Doggerland, a mythical landmass that is said to have connected the European mainland with the British Isles long time ago.

COZY LAND, 2018-19, follows three people in unicorn onesies who are carrying a doll – also dressed in a onesie – through a desolate coastal landscape. In what looks like a crossover between a consumerist bachelor party and an archaic ritual sacrifice procession, they move towards the Eastern Scheldt Storm Surge Barrier. Meanwhile, the text is read out as a voiceover.

References

Amechi, E. & Oni, B. (2019) 'Import of Electronic Waste into Nigeria: the Imperative of a Regulatory Policy Shift.' *Chinese Journal of Environmental Law*, 3(2), 141-166

Anonymous (1917) 'A German Deserter's War Experience.' Translated by J. Koettgen & B. W. Huebsch

Anson, P and Cummings, D. (1991) 'The First Space War: The contribution of satellites to the Gulf War.' *The RUSI Journal*, 136:4, 45-53

Barthes, R. (1972) *Mythologies*. Paladin

Bauman, Z. (2000) *Liquid Modernity*. Polity Press

Beauvoir, S. de (1953 [1949]) *The Second Sex*. Knopf

Beck, J. (2001) 'Without Form and Void: The American Desert as Trope and Terrain.' *Nepantla: Views from South 2*(1), 63-83

Beck, U. (1992) 'From Industrial Society to the Risk Society: Questions of Survival, Social Structure and Ecological Enlightenment.' *Theory, Culture & Society*, 9: 97-123

Bessel, R. (2015) *Violence: A modern obsession*. Simon & Schuster

Boulden, J. (2004) *CNN.com – Mobiles Used in High-Tech Terror – Apr 4*. [online] edition.cnn.com. <http://bit.ly/ploeger01>

Burleigh, T.J., Schoenherr, J.R. & Lacroix, G.L. (2013) 'Does the uncanny valley exist? An empirical test of the relationship between eeriness and the human likeness of digitally created faces.' *Computers in Human Behavior*, 29, 759-771

Carnahan, J. (2010) *The A-Team* [film]. <http://bit.ly/Ploeger15>

Casey, E.S. (1997) *The Fate of Place: A Philosophical History.* Univ. of California Press

Caslav Covino, D. (2004) *Amending the Abject Body: Aesthetic Makeovers in Medicine and Culture.* SUNY Press

Central Office of Information for Home Office (1975) *Protect and Survive* [video]. <http://bit.ly/Ploeger16>

Certeau, M. de (1984) *The Practice of Everyday Life.* Univ. of California Press

Chamayou, G. (2015 [2013]). *Drone Theory.* Penguin

Clark, R.P. & Cox, R. N. (1973) 'The Generation of Aerosols from the Human Body' in *Airborne Transmission and Airborne Infection.* Eds. J.F.P.H. Hers & K.C.Winkler, Oosthoek, 413-26

Davies, R. (2009) 'Meet the New Schtick.' <http://bit.ly/Ploeger12>

Deseret News (1991) 'Army Tanks Buried Iraqi Soldiers Alive in Trenches.' < http://bit.ly/Ploeger19 >

Dikkenberg, B v/d, (2012) 'Deltawerken goed voor internationale imago.' *Digibron.* <https://bit.ly/39rzRoo>

Douglas, M. (2002 [1966]) *Purity and Danger: An Analysis of Concepts of Pollution and Taboo.* Routledge and Keegan Paul

Dunifon, R., Duncan, G.J. & Brooks-Gunn, J. (2001) 'As Ye Sweep, So Shall Ye Reap'. *American Economic Review,* 91 (2): 150-154

Fine, G.A. & Hallett, T. (2003) 'Dust: A Study in Sociological Miniaturism.' *The Sociological Quarterly,* 44 (1): 1-15

Forge, S. (2007) 'Powering Down: Remedies for Unsustainable ICT.' *Foresight* 9(4), 3-21

Foucault, M. (1977 [1975]) *Discipline and Punish.* Pantheon Books

Gallagher, J.P. (1991) 'What Took You So Long?' in Assoc. of the US Army, *Personal Perspectives on the Gulf War,* 60-62

General Atomics (2012) *Predator C Avenger* [promotional video] <http://bit.ly/Ploeger17>

Gray, J. (1999) 'The Myth of Progress'. *New Statesman* <http://bit.ly/Ploeger36>

Haraway, D. (1985) 'Manifesto for Cyborgs: Science, technology, and socialist feminism in the 1980s.' *Socialist Review*, 80: 65-108

Hayles, N.K. (1999) *How We Became Posthuman, Virtual Bodies in Cybernetics, Literature, and Informatics*. Univ. of Chicago Press

Jentsch, E. (1997 [1906]), 'On the psychology of the uncanny', *Angelaki: Journal of the Theoretical Humanities*, 2 (1), 7-16

Kaplan, C. (2006) 'Precision Targets: GPS and the Militarization of US Consumer Identity.' *American Quarterly*, 58(3), 693-714

Kaplan, C., Loyer, E. & Daniels, E. (2013) 'Precision Targets: GPS and the Militarization of Everyday Life.' *Canadian Journal of Communication*, 38(3), 397-420

Katz, J. (2011) 'Xerox Phaser Drum Unit Hacked, Lives to Print Another Day.' <http://bit.ly/Ploeger03>

Keefe, J. R. (2009) 'The American Military and the Press: From Vietnam to Iraq', *Inquiries Journal*, 1 (10) <http://bit.ly/Ploeger13>

Koolhaas, R. (2002) 'Junkspace'. *October*, 100, 175-190

Kristeva, J. (1982) *Powers of Horror*. Columbia Univ. Press

Latour, B. (1999) *Pandora's Hope: Essays on the Reality of Science Studies*. Harvard Univ. Press

Lawrence of Arabia. (1962) [film] Dir. D. Lean. Columbia

Lefebvre, H. (1987) 'The Everyday and Everydayness.' *Yale French Studies*, (73), 7

_____ (1991[1947]) *Critique of Everyday Life: Vol. I*. Verso

Levinas, E., (2013 [1961]) *Totality and Infinity*. Duquesne Univ. Press

London, B. (1932) *Ending the Depression Through Planned Obsolescence* < http://bit.ly/Ploeger04>

MacDorman, K. F., & Ishiguro, H. (2006). 'The uncanny advantage of using androids in cognitive and social science research.' *Interaction Studies*, 7(3), 297-337

Marx, K. (1974 [1867]) *Capital (Vol. I)* Lawrence and Wishart

McGinn, B. (1994) 'Ocean and Desert as Symbols of Mystical Absorption in the Christian Tradition'. *The Journal of Religion*, 74(2), 155-181

McLuhan, M. (1964) *Understanding Media*. McGraw-Hill

McMaster, H.R. (1991) The Battle of 73 Easting <http://bit.ly/Ploeger26>

Mills, J. (2020) 'Police use drone to shame people for breaking coronavirus lockdown rules'. *Metro*, 26 March. <http://bit.ly/Ploeger27>

Mitchell, W. J., Szerszen, K., Sr., Lu, A. S., Schermerhorn, P. W., Scheutz, M. & MacDorman, K. F. (2011) 'A mismatch in the human realism of face and voice produces an uncanny valley'. *i-Perception*, 2(1), 10-12

Mori, M. (2005 [1970]), 'The Uncanny Valley', tr. Karl F. MacDorman and Takashi Minato. <http://bit.ly/Ploeger07>

No Country for Old Men. (2007) [film] Dir. J. & E. Coen. Paramount

Oxford Dictionary of English (2010) 3rd ed. Oxford Univ. Press

Patel, R. & Moore, J.W. (2017) *A History of the World in Seven Cheap Things: A guide to capitalism, nature, and the future of the planet*. Univ. of California Press

plapurdue (2008) Captive Penetrating Bolt Gun Cattle Stunner. <http://bit.ly/Ploeger28>

Poster, M. (2004) 'Consumption and Digital Commodities in the Everyday', *Cultural Studies*, 18:2-3, 409-423

Raiders of the Lost Ark (1981) [film] Dir. S. Spielberg. Paramount

rstahl, (2013) Smart Bomb/Missile Footage 1991. [video] <http://bit.ly/Ploeger20>

Said, E. (1978) *Orientalism*. Pantheon Books

Saygin, A.P., Chaminade, T., Ishiguro, H., Driver, J. & Frith, C. (2011) 'The Thing that Should Not Be: Predictive coding and the uncanny valley in perceiving human and humanoid robot actions.' *Social Cognitive and Affective Neuroscience*, 7(4), 413-22

Scanlan, J. (2005) *On Garbage*. Reaktion Books

Scarry, E. (1985) *The Body in Pain: The Making and Unmaking of the World*. Oxford Univ. Press

Seyama, J. & Nagayama, R. S. (2007) 'The Uncanny Valley: Effect of realism on the impression of artificial human faces'. *Presence: Teleoperators and Virtual Environments*, 16(4), 337-351

Shohat, E. (1997) 'Gender and Culture of Empire: Toward a Feminist Ethnography of the Cinema' in: M. Berstein & G. Studlar, *Visions Of The East: Orientalism in Film*. Rutgers Univ. Press

Sikkema, A. (2014) 'Zeeland is ingepolderd zonder poldermodel.' Resource, 24 September. <http://bit.ly/Ploeger37>

Sloyan, P. J. (1991) 'Iraqis Buried Alive - U.S. Attacked with Bulldozers During Gulf War Ground Attack'. *Seattle Times* <http://bit.ly/Ploeger21>

_____ (2003) 'What I saw was a bunch of filled-in trenches with people's arms and legs sticking out of them. For all I know, we could have killed thousands'. *The Guardian* <http://bit.ly/Ploeger22>

Sontag, S. (1998) [1977] 'On Photography' in Crowley, D. & Heyer, P., eds., *Communication in History: Technology, Culture, Society*, 3rd ed. Longman

Stahl, R. (2018) *Through the Crosshairs*. Rutgers Univ. Press

Steenhuis, M., ed. (2016) *De Deltawerken*. Nai10 Uitgevers

The News (2011) 'Karachi's ethnic composition undergoing radical change', 1 November <http://bit.ly/Ploeger18>

Tsoneva, J. (2019) 'The Revolt in the Trenches'. Jacobin, p.n.p. <http://bit.ly/Ploeger23>

Van Dyke, J.C. (1901) *The Desert: Further Studies in Natural Appearances*. Sampson Low, Marston

Veblen, T. (1899) *The Theory of The Leisure Class: An Economic Study Of Institutions*. Macmillan

Virilio, P. (2012) *The Administration of Fear*. Semiotext(e)

Visker, R. (2014) 'The Inhuman Core of Human Dignity: Levinas and Beyond'. *Levinas Studies*, 9, 1-21

Weber, M. (1958 [1905]) *The Protestant Eethic and the Spirit of Capitalism*. Scribner

Weiser, M. (1996) 'Ubiquitous Computing.' <http://bit.ly/Ploeger14>

Williams, B. G. (2010) 'The CIA's Covert Predator Drone War in Pakistan, 2004–2010: The History of an Assassination Campaign'. *Studies in Conflict & Terrorism*, 33:10, 871-892

Wolbring, G. (2008) 'The Politics of Ableism', *Development*, 51(2), 252-258

Wolfe, C. (2010) *What is Posthumanism?* Univ. of Minnesota Press

World Bank (2013) documents.worldbank.org: <http://bit.ly/Ploeger02>

Zeschky, M., Widenmayer, B. & Gassmann, O. (2011) 'Frugal Innovation in Emerging Markets', *Research-Technology Management*, 54:4, 38-45

About the author

Dani Ploeger is an artist and cultural theorist who explores situations of conflict and crisis on the fringes of the world of high-tech consumerism.

His artwork has been exhibited in museums, galleries and festivals worldwide, including ZKM Karlsruhe, Venice Architecture Biennale, transmediale, Nairobi National Museum and The New Institute in Rotterdam.

He holds a PhD from the University of Sussex, UK, and is currently a Research Fellow at The Royal Central School of Speech and Drama, University of London and a Fellow at V2_Lab for the unstable media in Rotterdam. His artwork is represented by Art Claims Impulse in Berlin.

www.daniploeger.org

About the Publisher

Triarchy Press is an independent publisher of books that bring a wider, systemic or contextual approach to many different areas of life, including:

- Government, Education, Health and other public services
- Ecology, Sustainability and Regenerative Cultures
- Leading and Managing Organisations
- The Money System
- Psychotherapy and Arts and other Expressive Therapies
- Walking, Psychogeography and Mythogeography
- Movement and Somatics
- Innovation
- The Future and Future Studies

For books by Nora Bateson, Daniel Wahl, Russ Ackoff, Barry Oshry, John Seddon, Phil Smith, Bill Tate, Patricia Lustig, Sandra Reeve, Graham Leicester, Nelisha Wickremasinghe, Bill Sharpe, Alyson Hallett and other remarkable writers, visit:

www.triarchypress.net